Proceedings of the
East Asia Joint Symposium
on Fields and Strings 2021

Proceedings of the
East Asia Joint Symposium on Fields and Strings 2021

Osaka City University, 22 – 27 November 2021

Editors

Satoshi Iso
KEK

Hiroshi Itoyama
Osaka City University, Japan

Kazunobu Maruyoshi
Seikei University, Japan

Takahiro Nishinaka
Osaka City University, Japan

Takeshi Oota
Osaka City University, Japan

Kazuhiro Sakai
Meiji Gakuin University, Japan

Asato Tsuchiya
Shizuoka University, Japan

Reiji Yoshioka
Osaka City University, Japan

NEW JERSEY · LONDON · SINGAPORE · BEIJING · SHANGHAI · HONG KONG · TAIPEI · CHENNAI · TOKYO

Published by

World Scientific Publishing Co. Pte. Ltd.

5 Toh Tuck Link, Singapore 596224

USA office: 27 Warren Street, Suite 401-402, Hackensack, NJ 07601

UK office: 57 Shelton Street, Covent Garden, London WC2H 9HE

British Library Cataloguing-in-Publication Data
A catalogue record for this book is available from the British Library.

ISBN 978-981-126-162-6 (hardcover)
ISBN 978-981-126-163-3 (ebook for institutions)
ISBN 978-981-126-164-0 (ebook for individuals)

For any available supplementary material, please visit
https://www.worldscientific.com/worldscibooks/10.1142/13004#t=suppl

Typeset by Stallion Press
Email: enquiries@stallionpress.com

Printed in Singapore

Preface

This volume contains the Proceedings of The East Asia Joint Symposium on Fields and Strings 2021, which was held in hybrid form at the Media Center of Osaka City University on November 22–27, 2021. About 160 physicists from all over East Asia attended or stayed online for this symposium and more than 50 researchers presented their results in the invited lectures, the short talks or the poster session. Quantum field theory and string theory in the context of several exciting developments were discussed, which include frontiers of supersymmetric gauge theory, anomalies and higher form symmetries and several issues on quantum gravity and black holes.

We thank all of the speakers and the participants of this symposium for their stimulating lectures and intensive discussions. It is our sincerest hope that this volume will not only help to advance our field by presenting the latest developments in research, but also to serve to inspire new generations of physicists.

The organizers:

Satoshi Iso
Hiroshi Itoyama (Chair)
Kazunobu Maruyoshi
Takahiro Nishinaka
Takeshi Oota
Kazuhiro Sakai
Asato Tsuchiya
Reiji Yoshioka

Contents

Ginzburg-Landau effective action for a fluctuating holographic superconductor

Yanyan Bu

School of Physics, Harbin Institute of Technology,
Harbin, 150001, China
E-mail: yybu@hit.edu.cn

Mitsutoshi Fujita

School of Physics and Astronomy, Sun Yat-Sen University,
Zhuhai, 519082, China
E-mail: fujita@mail.sysu.edu.cn

Shu Lin

School of Physics and Astronomy, Sun Yat-Sen University,
Zhuhai, 519082, China,
Guandong Provincial Key Laboratory of Quantum Metrology and Sensing, Sun
Yat-Sen University,
Zhuhai, 519082, China
E-mail: linshu8@mail.sysu.edu.cn

The fluctuation effect of the order parameter in a holographic superconductor model was analyzed under holographic prescription for Schwinger-Keldysh closed time contour for non-equilibrium system in the paper [1]. The time-dependent Ginzburg-Landau effective action, which governs the dynamics of the fluctuating order parameter near the critical point, is derived. The time-dependent Ginzburg-Landau action is computed up to quartic order of the fluctuating order parameter and first order in time derivative in a semi-analytical approach.

Keywords: Black holes, Gauge-gravity correspondence, Holography and condensed matter physics

1. Introduction

Critical exponents describe scaling behavior of observables near a critical point of continuous phase transitions. The characteristic properties of critical exponents are symmetry, dimension, and properties of order parameters. It is believed to be independent of the details of interaction. As a result, it is important to examine the critical region. The results of weak and strong coupling can be compared. Holography will make it possible to analyze dynamics (e.g. changing temperature) in the critical region: the critical exponent and the specific heat in theories at a quantum critical point. The gauge/gravity correspondence will give insights of cuprate superconductors, which appear in phase structure of superconductivity vs quantum criticality in interesting strongly correlated systems.

Holographic superconductors have become a driving force for establishing AdS/CMT (condensed matter theory). Holographic s-wave superconductors are referred to Abelian Higgs models[2,3], where s-wave, p-wave, and so on refer to the spatial component of the Cooper pair's wave function. The phase transition of an s-wave superconductor is from a black hole with no hair (normal phase) to one with scalar hair. The gap of the AC conductivity at $T = 0$ represents the frequency $\omega \sim 8T_c$. This is consistent with the experimental results of cuprate superconductors. Because it is different from the BCS prediction $\omega \sim 3.5T_c$, the holographic result implies the strong coupling limit. The paper[4] constructed the general off-shell Ginzburg-Landau formulation (analytic expressions) of a holographic superconductor. An inhomogeneous order parameter was used to calculate the scaling coefficient and gradient term. Holographic entanglement entropy has become a nice probe of the holographic superconductor phase transition first realized by papers[5-7].

In this work, we apply the holographic Schwinger-Keldysh approach[8] to a holographic superconductor model[9]. The time-dependent physics of a fluctuating order parameter is investigated. This model is applicable to nonlinear problems. The main goal is to introduce time-dependence in the Landau-Ginzburg action and to consider a non-equilibrium QFT framework.

2. The time-dependent Ginzburg-Landau effective action

In this section, we derive the time-dependent Ginzburg-Landau effective action from a holographic superconductor model. We analyze it in the probe limit without considering the backreaction: the scalar QED in AdS Schwarzschild geometry[10]. Consider an analytic holographic superconductor, the lagrangian of which also becomes the scalar QED as follows:

$$S_0 = \int d^5 x \sqrt{-g} \left[-\frac{1}{4} f_{\mu\nu} f^{\mu\nu} - (D_\mu \psi)^* (D^\mu \psi) - m_0^2 \psi^* \psi \right], \qquad (1)$$

$$m_0^2 = -4, \qquad (2)$$

where $f_{MN} = \nabla_\mu a_\nu - \nabla_\nu a_\mu$ and $D_\mu = \nabla_\mu - iqa_\mu$. $U(1)$ gauge symmetry in the bulk corresponds to $U(1)$ global symmetry on the AdS boundary. The formation of a charged scalar hair causes the spontaneous symmetry breaking of $U(1)$. The incoming Eddington-Finkelstein coordinate is used to parameterize the AdS Schwarzschild metric $x^\mu = (r, v, x^i)$.

Fig. 1. The radial coordinate is complexified in the gravity dual. The picture is taken from the paper[12]

To analyze both fluctuation and dissipation, the boundary theory must be placed on the Schwinger-Keldysh time contour[11]. A holographic dual has the complexified radial coordinate[8]. It circles the black hole horizon from infinity 1 to infinity 2. See Fig. 1.

The solution of the gauge field is in the leading order

$$a_v = \mu_0 \left(1 - \frac{r_h^2}{r^2}\right). \tag{3}$$

If the regular boundary condition at the black hole horizon is required, it is known that ODE can be solved analytically at the special point $\mu_0 = 2r_h$. This unique value corresponds to the critical point at which the phase transition occurs.

The scalar field has mass $m_0^2 = -4$ which saturates the Breitenlohner-Freedman bound. The AdS boundary expansion of ψ, ψ^* becomes

$$\psi(r \to \infty_s) = \psi_{bs} \frac{\log r}{r^2} + \frac{\Phi_s}{r^2} + \cdots,$$

$$\psi^*(r \to \infty_s) = \bar\psi_{bs} \frac{\log r}{r^2} + \frac{\bar\Phi_s}{r^2} + \cdots, \tag{4}$$

Because ψ, ψ^* are not independent, $\bar\psi_{bs}$ ($\bar\Phi_s$) are not always complex conjugate of ψ_{bs} (Φ_s). The two modes have conformal dimension two and are normalizable. The two modes are canonical conjugates of one another[13].

We fix Φ_s, identified as the expectation value of condensate, by adding a proper boundary terms at the AdS boundary.

The EOMs are nonlinear partial differential equations (PDEs). It is not easy to obtain an analytic solution for PDEs. Therefore, we consider the following perturbation scheme to simplify the problem in the high temperature phase where temperature T is slightly above T_c. We consider (1) off-equilibrium fluctuation of Φ while keeping Φ^2, Φ^4 close to the critical point. (2) A slight deviation from the critical point μ_0 is described by a chemical potential perturbation $O(\delta\mu)$, (3) We require the hydrodynamic limit: the system grows slowly with increase of time. The hydrodynamic limit can be represented as the low frequency expansion $O(\partial_v)$. Each a field is described by the triple expansion $A_v^{(l)(m)(n)}$, $\Psi^{(l)(m)(n)}$, $\Psi^{*(l)(m)(n)}$.

We compute the partially on-shell action, which leads to the boundary effective action S_{eff}. UV divergences are removed by counter-term action added at the AdS boundaries. Due to the triple expansion of the bulk fields, the effective action $S_{eff} = \int d^4x \mathcal{L}$ is expanded as

$$\mathcal{L} = \mathcal{L}^{(0)(2)(0)} + \mathcal{L}^{(0)(2)(1)} + \mathcal{L}^{(0)(4)(0)} + \mathcal{L}^{(1)(2)(0)} + \dots \tag{5}$$

Effective action in each order becomes

$$\mathcal{L}^{(0)(2)(0)} = -\frac{\Phi_1 - \Phi_2}{2i\pi}(\Phi_1^* - \Phi_2^*) = \frac{i}{2\pi}\Phi_a^*\Phi_a. \tag{6}$$

$$\mathcal{L}^{(0)(2)(1)} = \delta\mu\left[\frac{\log 2}{i\pi}(\Phi_2^* - \Phi_1^*)(\Phi_2 - \Phi_1) - (\Phi_2^*\Phi_2 - \Phi_1^*\Phi_1)\right]$$

$$= \delta\mu\left[\frac{\log 2}{i\pi}\Phi_a^*\Phi_a + (\Phi_a\Phi_r^* + \Phi_a^*\Phi_r)\right], \tag{7}$$

$$\mathcal{L}^{(1)(2)(0)} = -\frac{1}{4}(1 - 3i)\Phi_a^*\partial_v\Phi_r + \frac{1}{4}(1 + 3i)\Phi_r^*\partial_v\Phi_a + \frac{\log 2}{4\pi}\Phi_a^*\partial_v\Phi_a, \tag{8}$$

$$\mathcal{L}^{(0)(4)(0)} = -0.000129006i(\Phi_a\Phi_a^*)^2 + 0.00466688\Phi_a\Phi_a^*(\Phi_a^*\Phi_r + \Phi_a\Phi_r^*)$$

$$- 0.000263406i\left[(\Phi_a^*\Phi_r)^2 + (\Phi_a\Phi_r^*)^2\right] - 0.00105363i\Phi_a\Phi_r\Phi_a^*\Phi_r^*$$

$$+ 0.0208333(\Phi_a^*\Phi_r^*\Phi_r^2 + \Phi_a\Phi_r\Phi_r^{*2}), \tag{9}$$

where we have introduced difference and average combinations as the (r,a)-basis:

$$\Phi_a = \Phi_1 - \Phi_2, \quad \Phi_r = \frac{1}{2}(\Phi_1 + \Phi_2). \tag{10}$$

Equation (6) is the noise term. The coefficient should be a positive imaginary number because of unitarity. The chemical potential correction is

5

represented by equation (7). This also includes corrections of $\Phi_a^*\Phi_a$. Equation (8) is the first order of hydrodynamic corrections. These terms contribute the symmetric and retarded two-point correlations, which obey the standard fluctuation-dissipation theorem up to $\delta\mu$ at lowest order in ω. Equation (9) includes general 8 potential terms.

Effective action has Z_2-reflection symmetry $\Phi_1 \leftrightarrow \Phi_2$ and $\Phi_1^* \leftrightarrow \Phi_2^*$ with sign flips of effective action. This implies that effective action has complex coefficients according to [14]. Higher-point correlations will be constrained by Z_2 symmetry.

3. Comparison with weakly coupled results

We compare gravity dual with weakly coupled results. The time-dependent Ginzburg-Landau action S_{GL} $(T > T_c)$ [11] becomes

$$S_{\mathrm{GL}} = 2\nu \int d^4x \left[\Phi_{\mathcal{K}}^{\mathrm{q}*}(L^{-1})^R \Phi_{\mathcal{K}}^{\mathrm{cl}} + \Phi_{\mathcal{K}}^{\mathrm{cl}*}(L^{-1})^A \Phi_{\mathcal{K}}^{\mathrm{q}} + \Phi_{\mathcal{K}}^{\mathrm{q}*}(L^{-1})^K \Phi_{\mathcal{K}}^{\mathrm{q}} \right], \quad (11)$$

where $\vec{\Phi}_{\mathcal{K}} = (\vec{\Phi}_{\mathcal{K}}^{\mathrm{cl}}, \vec{\Phi}_{\mathcal{K}}^{\mathrm{q}})^{\mathrm{T}}$ are the classical part and quantum fluctuation of the scalar condensate (the order parameter). Matrix elements become

$$(L^{-1})^{R(A)} = \frac{\pi}{8T} \left[\mp\partial_t + D(\nabla_r + 2ieA_{\mathcal{K}}^{\mathrm{cl}})^2 - \tau_{\mathrm{GL}}^{-1} - \frac{7\zeta(3)}{\pi^3 T_c}|\Phi_{\mathcal{K}}^{\mathrm{cl}}|^2 \right], \quad (12)$$

$$(L^{-1})^K = \coth\frac{\omega}{2T} \left[(L^{-1})^R(\omega) - (L^{-1})^A(\omega) \right] \approx \frac{i\pi}{2}, \quad (13)$$

where signs in the time-derivative term change in $R(A)$ and $\tau_{\mathrm{GL}} = \pi/[8(T-T_c)]$. The symbol "$\mathcal{K}$" means the $\mathcal{K}$-gauge: the time-component of the external gauge potential is equal to zero. The equation (13) is obtained from Kubo-Martin-Schwinger symmetry.

We have two fundamental parameters in the action: ν—the density of states, and D—the diffusion constant.

We have a similarity between the holographic model and weakly coupled theory.

- The imaginary part of $\Phi_{\mathcal{K}}^{\mathrm{q}*}\Phi_{\mathcal{K}}^{\mathrm{q}}$ and eq. (6) are positive. This is a requirement coming from unitarity. In the gravity dual, the same term is corrected by eq. (7), which does not exist in the field theory side.
- the coefficient of the $\Phi_{\mathcal{K}}^{*cl}\Phi_{\mathcal{K}}^{q}$ (and $\Phi_{\mathcal{K}}^{*q}\Phi_{\mathcal{K}}^{cl}$) term contains inverse relaxation time, which shows $\tau_{GL} \sim \epsilon_T^{-z\nu} \sim \epsilon_T^{-1}$. Actually, both models are in model A of the Hohenberg-Halperin classification for dynamic universality class. Static and dynamic critical exponents

$\nu = 1/2$ and $z = 2$. The effective lagrangian eq. (7) in the gravity dual shows the critical exponent of the same dynamical universality class $\delta\mu \sim \mu - \mu_0 \sim T_c - T$. Critical exponents are calculated for holographic s-wave and p-wave superconductors[1,15].

- Only two quartic terms are present in eq. (11). These cover only $arrr$-terms. Eq. (9), on the other hand, has 8 general quartic potential terms. The gravity dual predicts more general results of holographic superconductors.

4. Discussion

We have outlined our obtained results[1]. We showed that holography provided a possible description of strongly coupled superconductors. In particular, time dependent Ginzburg-Landau effective action was derived from a holographic superconductor. Effective action is up to the quartic order in the fluctuating scalar condensate. This is an effective action for charge degrees of freedom such as currents and charge-condensate coupling. The holographic model was comparable with weak coupled BCS superconductor[11] and demonstrated the same dynamic universality class. Our work shows that the holographic Schwinger-Keldysh contour approach for non-equilibrium physics is relevant to nonlinear problems in the gravity dual.

The analysis in this work is for spatially homogeneous case, in which the charge diffusion part is decoupled. It would be interesting to consider spatially non-homogeneous case.

We are interested in how to incorporate Kibble-Zurek scaling. We must analyze the influence of the noise on the time dependent dynamics in quenching physics at that time.

References

1. Y. Bu, M. Fujita and S. Lin, Phys. Rev. D **101**, no.2, 026003 (2020) [arXiv:1906.00681 [hep-th]].
2. S. A. Hartnoll, C. P. Herzog and G. T. Horowitz, Phys. Rev. Lett. **101**, 031601 (2008) [arXiv:0803.3295 [hep-th]].
3. S. A. Hartnoll, C. P. Herzog and G. T. Horowitz, JHEP **12**, 015 (2008) [arXiv:0810.1563 [hep-th]].
4. L. Yin, D. Hou and H. c. Ren, Phys. Rev. D **91**, no. 2, 026003 (2015) [arXiv:1311.3847 [hep-th]].
5. T. Albash and C. V. Johnson, JHEP **05**, 079 (2012) [arXiv:1202.2605 [hep-th]].

6. R. G. Cai, S. He, L. Li and Y. L. Zhang, JHEP **07**, 088 (2012) doi:10.1007/JHEP07(2012)088 [arXiv:1203.6620 [hep-th]].

7. R. G. Cai, S. He, L. Li and Y. L. Zhang, JHEP **07**, 027 (2012) doi:10.1007/JHEP07(2012)027 [arXiv:1204.5962 [hep-th]].

8. P. Glorioso, M. Crossley and H. Liu, [arXiv:1812.08785 [hep-th]]. LaTeX (US)

9. C. P. Herzog, Phys. Rev. D **81**, 126009 (2010) [arXiv:1003.3278 [hep-th]].

10. S. S. Gubser, Phys. Rev. D **78**, 065034 (2008) [arXiv:0801.2977 [hep-th]].

11. A. Kamenev, Cambridge University Press, 2011

12. Y. Bu, M. Fujita and S. Lin, JHEP **09**, 168 (2021) [arXiv:2106.00556 [hep-th]].

13. I. R. Klebanov and E. Witten, Nucl. Phys. B **556**, 89-114 (1999) [arXiv:hep-th/9905104 [hep-th]].

14. M. Crossley, P. Glorioso and H. Liu, JHEP **09**, 095 (2017) [arXiv:1511.03646 [hep-th]].

15. K. Maeda, M. Natsuume and T. Okamura, Phys. Rev. D **79**, 126004 (2009) [arXiv:0904.1914 [hep-th]].

Complex Langevin studies of the emergent space–time in the type IIB matrix model

Kohta Hatakeyama[1], Konstantinos Anagnostopoulos[2], Takehiro Azuma[3], Mitsuaki Hirasawa[4], Yuta Ito[5], Jun Nishimura[1,6], Stratos Papadoudis[2], and Asato Tsuchiya[7]

[1] *Theory Center, Institute of Particle and Nuclear Studies, High Energy Accelerator Research Organization (KEK), 1-1 Oho, Tsukuba, Ibaraki 305-0801, Japan E-mail: khat@post.kek.jp, jnishi@post.kek.jp*

[2] *Physics Department, School of Applied Mathematical and Physical Sciences, National Technical University of Athens, Zografou Campus, GR-15780 Athens, Greece E-mail: konstant@mail.ntua.gr, sp10018@central.ntua.gr*

[3] *Setsunan University, 17-8 Ikeda Nakamachi, Neyagawa, Osaka, 572-8508, Japan E-mail: azuma@mpg.setsunan.ac.jp*

[4] *Sezione di Milano-Bicocca, Istituto Nazionale di Fisica Nucleare (INFN), Piazza della Scienza 3, I-20126 Milano, Italy E-mail: Mitsuaki.Hirasawa@mib.infn.it*

[5] *National Institute of Technology, Tokuyama College, Gakuendai, Shunan, Yamaguchi 745-8585, Japan E-mail: y-itou@tokuyama.ac.jp*

[6] *Department of Particle and Nuclear Physics, School of High Energy Accelerator Science, Graduate University for Advanced Studies (SOKENDAI), 1-1 Oho, Tsukuba, Ibaraki 305-0801, Japan*

[7] *Department of Physics, Shizuoka University, 836 Ohya, Suruga-ku, Shizuoka 422-8529, Japan E-mail: tsuchiya.asato@shizuoka.ac.jp*

The IIB matrix model has been proposed as a non-perturbative definition of superstring theory since 1996. We study a simplified model that describes the late time behavior of the IIB matrix model non-perturbatively using Monte Carlo methods, and we use the complex Langevin method to overcome the sign problem. We investigate a scenario where the space–time signature changes dynamically from Euclidean at early times to Lorentzian at late times.We discuss the possibility of the emergence of the (3+1)D expanding universe.

Keywords: Type IIB matrix model; Emergent space–time; Complex Langevin simulation

1. Introduction

Superstring theory is the most promising candidate for a unified theory of all interactions, including quantum gravity. The theory is consistently defined in ten-dimensional space-time, leading to the compacting of the extra dimensions into small compact internal spaces. These scenarios have been investigated perturbatively on D-brane backgrounds and result in a vast number of vacua, leading to the so-called string landscape. It is, therefore, interesting to see what happens when one includes non-perturbative effects and whether these play an essential role in determining the true vacuum of the theory. The type IIB matrix model [1] has been proposed as a non-perturbative definition of superstring theory and provides a promising context to study such problems.

The type IIB matrix model is formally obtained by the dimensional reduction of ten-dimensional $\mathcal{N} = 1$ Super Yang–Mills (SYM) to zero dimensions. The theory has maximal $\mathcal{N} = 2$ supersymmetry (SUSY), where translations are realized by the shifts $A_\mu \to A_\mu + \alpha_\mu \mathbf{1}$, $\mu = 0, \dots, 9$. The eigenvalues of the bosonic matrices A_μ can therefore be interpreted as coordinates of space–time. Thus, in this model, space–time appears dynamically from the degrees of freedom of matrices. In the Euclidean version of the model, the Spontaneous Symmetry Breaking (SSB) of the SO(10) rotational symmetry down to SO(3) occurs, which implies the emergence of a three-dimensional space [2–8].

By Monte Carlo simulation [9], it was found that a continuous time emerges dynamically, and a three-dimensional space expands. In Refs. [10, 11], it turned out that the expanding behavior of the space obeys the exponential law at early times and the power-law at late times. In Ref. [12], however, it was shown that SSB comes from singular configurations associated with the Pauli matrices, in which only two eigenvalues are large. This problem has been attributed to an approximation used to avoid the sign problem, which turned out later to be unjustifiable.

In Refs. [13–15], the Complex Langevin Method (CLM) [16, 17] was used to overcome the sign problem without the above mentioned approximation. When one applies this method, one should apply the criterion for correct convergence of the CLM [18–24]. In Ref. [15], we found a new phase in which the structure of space is continuous by applying the CLM to the Lorentzian type IIB matrix model. See also Refs. [25–27] for other related works.

In this work, we study the bosonic version of the type IIB matrix model by using the CLM. We show the equivalence between the Lorentzian and Euclidean models, which implies that the space–time in the Lorentzian model is Euclidean. To realize the possibility of the dynamical change of signature from

Euclidean to Lorentzian, we introduce a Lorentz-invariant mass term in the action that breaks the equivalence. We find some evidence that the signature of space–time changes from Euclidean at early times to Lorentzian at later times.

2. The type IIB matrix model

2.1. Definition

The action of the type IIB matrix model is given as follows: $S = S_{\rm b} + S_{\rm f}$,

$$S_{\rm b} = -\frac{1}{4g^2} \operatorname{Tr}\left([A^\mu, A^\nu][A_\mu, A_\nu]\right) , \quad S_{\rm f} = -\frac{1}{2g^2} \operatorname{Tr}\left(\bar{\Psi}(\mathcal{C}\Gamma^\mu)[A_\mu, \Psi]\right) , \quad (1)$$

where A_μ ($\mu = 0, \ldots, 9$) and Ψ are $N \times N$ Hermitian matrices, and Γ^μ and $\mathcal{C}$ are 10-dimensional gamma matrices and the charge conjugation matrix, respectively, which are obtained after the Weyl projection. The A_μ and Ψ transform as vectors and Majorana-Weyl spinors under SO(9,1) transformations. In this study, we omit $S_{\rm f}$ to reduce the computational cost.

The partition function is given by $Z = \int dA e^{iS_{\rm b}}$. Due to the phase factor $e^{iS_{\rm b}}$, the model is not well-defined as it is, and in this work, we define it by deforming the integration contour. When we rewrite the partition function as $Z = \int dA e^{-\tilde{S}}$, the action of the Lorentzian model is given as

$$\tilde{S} = -\frac{i}{4}N\left[-2\operatorname{Tr}(F_{0i})^2 + \operatorname{Tr}(F_{ij})^2\right] , \quad (2)$$

where $g^2 = 1/N$ and $F_{\mu\nu} = i[A_\mu, A_\nu]$. According to Cauchy's theorem, one can rotate the Lorentzian matrices A_μ to the Euclidean ones $\tilde{A}_\mu$ since the integration contour of A_μ can be deformed keeping the real part of $\tilde{S}$ positive. The relationship between A_μ and $\tilde{A}_\mu$ is

$$A_0 = e^{-i\frac{3\pi}{8}}\tilde{A}_0 , \quad A_i = e^{i\frac{\pi}{8}}\tilde{A}_i . \quad (3)$$

Then, the Euclidean action is given by

$$\tilde{S} = \frac{1}{4}N\left[2\operatorname{Tr}(\tilde{F}_{0i})^2 + \operatorname{Tr}(\tilde{F}_{ij})^2\right] , \quad (4)$$

which is positive-definite. Here we have defined $\tilde{F}_{\mu\nu} = i[\tilde{A}_\mu, \tilde{A}_\nu]$.

2.2. Equivalence between the Euclidean and Lorentzian models

By using Eq. (3), one can derive the relationship between the expectation values of $\operatorname{Tr} A_0^2$ and $\operatorname{Tr} A_i^2$ in the two models:

$$\left\langle \frac{1}{N}\operatorname{Tr} A_0^2 \right\rangle_{\rm L} = e^{-i\frac{3\pi}{4}}\left\langle \frac{1}{N}\operatorname{Tr} \tilde{A}_0^2 \right\rangle_{\rm E} , \quad \left\langle \frac{1}{N}\operatorname{Tr} A_i^2 \right\rangle_{\rm L} = e^{i\frac{\pi}{4}}\left\langle \frac{1}{N}\operatorname{Tr} \tilde{A}_i^2 \right\rangle_{\rm E} , \quad (5)$$

12

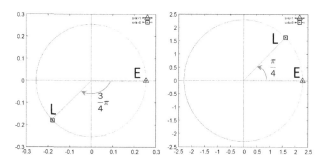

Fig. 1. We plot the expectation values of $\frac{1}{N}\operatorname{Tr}A_0^2$ (Left) and $\frac{1}{N}\operatorname{Tr}A_i^2$ (Right). Those of the Euclidean and Lorentzian models are represented by the triangles and the squares, respectively. The angles between the Lorentzian and Euclidean models of $\langle\frac{1}{N}\operatorname{Tr}\tilde{A}_0^2\rangle$ and $\langle\frac{1}{N}\operatorname{Tr}\tilde{A}_i^2\rangle$ are $-3\pi/4$ and $\pi/4$, which agree with Eq. (5).

where $\langle\,\cdot\,\rangle_L$ and $\langle\,\cdot\,\rangle_E$ denote the expectation values in the Lorentzian and Euclidean models, respectively. In Fig. 1 (Left), $\langle\frac{1}{N}\operatorname{Tr}A_0^2\rangle$ is shown, and the angle between $\langle\frac{1}{N}\operatorname{Tr}\tilde{A}_0^2\rangle_E$ and $\langle\frac{1}{N}\operatorname{Tr}A_0^2\rangle_L$ is $-3\pi/4$. In Fig. 1 (Right), $\langle\frac{1}{N}\operatorname{Tr}A_i^2\rangle$ is shown, and the angle between $\langle\frac{1}{N}\operatorname{Tr}\tilde{A}_i^2\rangle_E$ and $\langle\frac{1}{N}\operatorname{Tr}A_i^2\rangle_L$ is $\pi/4$. These angles are in agreement with Eq. (5).

These results are consistent with the fact that the Lorentzian and the Euclidean models are equivalent. Expectation values in the Lorentzian model can be obtained by simply rotating the phase of those in the Euclidean model. In particular, $\langle\frac{1}{N}\operatorname{Tr}A_0^2\rangle_L$ and $\langle\frac{1}{N}\operatorname{Tr}A_i^2\rangle_L$ are complex and the emergent space–time should be interpreted as Euclidean.

2.3. Lorentz-invariant mass term

To realize real time and space, we introduce a *Lorentz-invariant mass term* in the action. For the Lorentzian model, the action is

$$\tilde{S} = -\frac{i}{4}N\left[-2\operatorname{Tr}(F_{0i})^2 + \operatorname{Tr}(F_{ij})^2\right] - \frac{i}{2}N\gamma\left[\operatorname{Tr}(A_0)^2 - \operatorname{Tr}(A_i)^2\right] \quad (6)$$

with $\gamma > 0$. Using Eq. (3), we find that the action for the corresponding Euclidean model becomes

$$\tilde{S} = \frac{1}{4}N\left[2\operatorname{Tr}(\tilde{F}_{0i})^2 + \operatorname{Tr}(\tilde{F}_{ij})^2\right] + \frac{1}{2}N\gamma\,e^{i\frac{3\pi}{4}}\left[\operatorname{Tr}(\tilde{A}_0)^2 + \operatorname{Tr}(\tilde{A}_i)^2\right], \quad (7)$$

where the real part of the mass term is negative. If $\gamma < 0$, the real part of the mass term in the Euclidean model becomes positive, and then the matrices can be rotated from the Lorentzian to the Euclidean, which implies the equivalence between the two models.

The same mass term was used to study classical solutions of the Lorentzian type IIB matrix model [28]:

$$[A^\nu, [A_\nu, A_\mu]] - \gamma A_\mu = 0 . \tag{8}$$

For $\gamma > 0$, one can obtain classical solutions with smooth space and expanding behavior. Classical solutions with Hermitian A_μ make the time and space real. For $\gamma = 0$, the classical solutions are given by simultaneously diagonalizable A_μ, which do not necessarily have an expanding behavior. For $\gamma < 0$, there do not exist classical solutions with expanding behavior.

2.4. The time evolution

As mentioned in Sec. 1, time does not exist a priori, and we define it as follows. We choose a basis in which A_0 is diagonal and its eigenvalues are in the ascending order: $A_0 = \mathrm{diag}(\alpha_1, \alpha_2, \ldots, \alpha_N)$, $\alpha_1 \le \alpha_2 \le \cdots \le \alpha_N$. Then, we define $\bar\alpha_k$ as $\bar\alpha_k = \frac{1}{n} \sum_{i=1}^{n} \alpha_{k+i}$, and the time t_ρ as

$$t_\rho = \sum_{k=1}^{\rho} |\bar\alpha_{k+1} - \bar\alpha_k| . \tag{9}$$

Here, we introduce the $n \times n$ matrices $\bar A_i(t)$ as $(\bar A_i)_{ab}(t) = (A_i)_{k+a, k+b}$, which represent the space at the time t.

3. Complex Langevin method

The complex Langevin method (CLM) [16, 17] can be applied successfully to many systems with a complex action problem. One writes down stochastic differential equations for the complexified degrees of freedom, which can be used to compute expectation values under certain conditions. Consider a model given by the partition function $Z = \int dx\, w(x)$, where $x \in \mathbb{R}^n$ and $w(x)$ is a complex-valued function. In the CLM, we complexify the variables $x \in \mathbb{R}^n \longrightarrow z \in \mathbb{C}^n$, and solve the complex Langevin equation with the Langevin time σ:

$$\frac{dz_k}{d\sigma} = \frac{1}{w(z)} \frac{\partial w(z)}{\partial z_k} + \eta_k(\sigma) . \tag{10}$$

The first term of the right-hand side of Eq. (10) is the drift term, and the second one is the real Gaussian noise with the probability distribution

$$\mathrm{P}(\eta_k(\sigma)) \propto e^{-\frac{1}{4} \int d\sigma \sum_k [\eta_k(\sigma)]^2} . \tag{11}$$

To confirm that the CLM gives correct solutions, we use the criterion that the probability distribution of the drift term should be exponentially suppressed for large values [23].

3.1. *Application of the CLM to the type IIB matrix model*

To apply the CLM to the type IIB matrix model, we make a change of variables [13]: $\alpha_1 = 0$, $\alpha_i = \sum_{k=1}^{i-1} e^{\tau_k}$ for $2 \le i \le N$, where we introduce new real variables τ_k. In this way, the ordering of α_i is automatically realized. Initially, α_i are real, and A_i are Hermitian matrices. To apply the CLM, we complexify τ_k and take A_i to be SL$(N, \mathbb{C})$ matrices. The complex Langevin equations are given by

$$\frac{d\tau_k}{d\sigma} = -\frac{\partial S_{\text{eff}}}{\partial \tau_k} + \eta_k(\sigma) , \quad \frac{d(A_i)_{kl}}{d\sigma} = -\frac{\partial S_{\text{eff}}}{\partial (A_i)_{lk}} + (\eta_i)_{kl}(\sigma) , \quad (12)$$

where S_{eff} is obtained from $\tilde{S}$ in Eq. (6) by adding a term associated with the gauge fixing and the Jacobian term associated with the change of variables.

4. Results

In the following, we introduce a parameter ε in the mass term:

$$\tilde{S} = -\frac{i}{4}N\left[-2\operatorname{Tr}(F_{0i})^2 + \operatorname{Tr}(F_{ij})^2\right] - \frac{i}{2}N\gamma\left[e^{i\varepsilon}\operatorname{Tr}(A_0)^2 - e^{-i\varepsilon}\operatorname{Tr}(A_i)^2\right] \quad (13)$$

to shift coefficients of $\operatorname{Tr}(A_0)^2$ and $\operatorname{Tr}(A_i)^2$ slightly from pure imaginary, and set $\varepsilon = \pi/10$.

4.1. *Expectation value of the time coordinate*

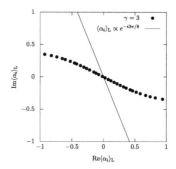

Fig. 2. Expectation values of the eigenvalues α_i of A_0 for $N = 32, \gamma = 3$ are plotted. The solid line corresponds to the Euclidean model, where the complex phase of the expectation values $\langle \alpha_i \rangle_{\text{L}}$ is $\exp(-i3\pi/8)$. From this plot, θ_{t} tends to become 0 at late times (at both ends of the distribution).

When $\gamma = 0$, Eq. (3) holds, and we expect that $\langle \alpha_i \rangle_{\text{L}} = e^{-i\frac{3\pi}{8}} \langle \tilde{\alpha}_i \rangle_{\text{E}}$. This is true because of the equivalence between the Euclidean and Lorentzian models, and time is regarded as the Euclidean one. We measure the time differences $(\Delta\alpha_i)_\mathrm{L} = (\alpha_{i+1})_\mathrm{L} - (\alpha_i)_\mathrm{L} \propto e^{i\theta_\mathrm{t}}$. The emergent time is real if $\theta_\mathrm{t} = 0$.

In Fig. 2, we plot the expectation values of the time coordinates $\langle\alpha_i\rangle_\mathrm{L}$ on the complex plane for $N = 32, \gamma = 3$. The solid line corresponds to the Euclidean model, where the complex phase of $\langle\alpha_i\rangle_\mathrm{L}$ is $\exp(-i3\pi/8)$. From the plot, θ_t tends to become 0 at late times (at both ends of the distribution).

4.2. Time evolution of space

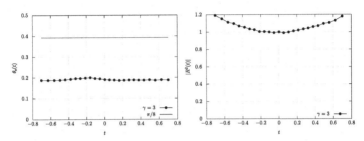

Fig. 3. (Left) $\theta_\mathrm{s}(t)$ is plotted against t for $\gamma = 3$. All values of $\theta_\mathrm{s}(t)$ are about 0.2 and below the $\theta_\mathrm{s}(t) = \pi/8$ line, which corresponds to the Euclidean space. (Right) $|R^2(t)|$ is plotted against t for $\gamma = 3$. We can see that the space is expanding slightly with the time t.

The time evolution of the extent of space is given by $R^2(t) = \left\langle \frac{1}{n}\mathrm{tr}\left(\bar{A}_i(t)\right)^2 \right\rangle = e^{2i\theta_\mathrm{s}(t)}|R^2(t)|$. Since the matrices $\bar{A}_i$ are complex, $R^2(t)$ is also complex. The time t is defined in Eq. (9). From Eq. (5), we obtain the Euclidean space when $\theta_\mathrm{s}(t) \sim \pi/8$, and the real space in the Lorentzian model when $\theta_\mathrm{s}(t) \sim 0$. Therefore, the signature of space–time can change dynamically in this model.

In Fig. 3 (Left) and (Right), $\theta_\mathrm{s}(t)$ and $|R^2(t)|$ are plotted against t for $N = 32, \gamma = 3$, respectively. All values of $\theta_\mathrm{s}(t)$ are about 0.2 and below the $\theta_\mathrm{s}(t) = \pi/8$ line, which corresponds to the Euclidean space. We can see that the space is expanding slightly with the time t from the plot of $|R^2(t)|$.

5. Conclusions

In this work, the CLM was applied to the bosonic type IIB matrix model in order to overcome the sign problem. We showed that the Lorentzian and Euclidean models are equivalent and that expectation values in the two models are related to each other by some complex phase rotation. The expectation values (5) in the Lorentzian model are complex, and space–time is Euclidean.

We introduced the model with a Lorentz-invariant mass term, which is a promising way to realize real time and expanding space. Then, the Euclidean and Lorentzian models are not equivalent anymore for $\gamma > 0$. We found that the time, which is extracted from the expectation values of the eigenvalues of A_0 in the Lorentzian model, may be real at late times although they are complex near the origin. We also studied the evolution of the extent of space with time. We have seen some tendency that the space becomes closer to real than the original model.

To obtain a three-dimensional expanding space, we expect that supersymmetry will play an essential role. We are currently investigating its effect, which we will report in the near future.

Acknowledgments

T. A., K. H. and A. T. were supported in part by Grant-in-Aid (Nos. 17K05425, 19J10002, and 18K03614, 21K03532, respectively) from Japan Society for the Promotion of Science. This research was supported by MEXT as "Program for Promoting Researches on the Supercomputer Fugaku" (Simulation for basic science: from fundamental laws of particles to creation of nuclei, JPMXP1020200105) and JICFuS. This work used computational resources of supercomputer Fugaku provided by the RIKEN Center for Computational Science (Project ID: hp210165) and Oakbridge-CX provided by the University of Tokyo (Project IDs: hp200106, hp200130, hp210094) through the HPCI System Research Project. Numerical computation was also carried out on PC clusters in KEK Computing Research Center. This work was also supported by computational time granted by the Greek Research and Technology Network (GRNET) in the National HPC facility ARIS, under the project IDs SUSYMM and SUSYMM2.

References

[1] N. Ishibashi, H. Kawai, Y. Kitazawa and A. Tsuchiya, A Large N reduced model as superstring, *Nucl. Phys.* **B498**, 467 (1997).

[2] J. Nishimura and F. Sugino, Dynamical generation of four-dimensional space-time in the IIB matrix model, *JHEP* **05**, p. 001 (2002).

[3] H. Kawai, S. Kawamoto, T. Kuroki, T. Matsuo and S. Shinohara, Mean field approximation of IIB matrix model and emergence of four-dimensional space-time, *Nucl. Phys. B* **647**, 153 (2002).

[4] T. Aoyama and H. Kawai, Higher order terms of improved mean field

approximation for IIB matrix model and emergence of four-dimensional space-time, *Prog. Theor. Phys.* **116**, 405 (2006).

[5] J. Nishimura, T. Okubo and F. Sugino, Systematic study of the SO(10) symmetry breaking vacua in the matrix model for type IIB superstrings, *JHEP* **10**, p. 135 (2011).

[6] K. N. Anagnostopoulos, T. Azuma and J. Nishimura, Monte Carlo studies of the spontaneous rotational symmetry breaking in dimensionally reduced super Yang-Mills models, *JHEP* **11**, p. 009 (2013).

[7] K. N. Anagnostopoulos, T. Azuma, Y. Ito, J. Nishimura and S. K. Papadoudis, Complex Langevin analysis of the spontaneous symmetry breaking in dimensionally reduced super Yang-Mills models, *JHEP* **02**, p. 151 (2018).

[8] K. N. Anagnostopoulos, T. Azuma, Y. Ito, J. Nishimura, T. Okubo and S. Kovalkov Papadoudis, Complex Langevin analysis of the spontaneous breaking of 10D rotational symmetry in the Euclidean IKKT matrix model, *JHEP* **06**, p. 069 (2020).

[9] S.-W. Kim, J. Nishimura and A. Tsuchiya, Expanding (3+1)-dimensional universe from a Lorentzian matrix model for superstring theory in (9+1)-dimensions, *Phys. Rev. Lett.* **108**, p. 011601 (2012).

[10] Y. Ito, S.-W. Kim, Y. Koizuka, J. Nishimura and A. Tsuchiya, A renormalization group method for studying the early universe in the Lorentzian IIB matrix model, *PTEP* **2014**, p. 083B01 (2014).

[11] Y. Ito, J. Nishimura and A. Tsuchiya, Power-law expansion of the Universe from the bosonic Lorentzian type IIB matrix model, *JHEP* **11**, p. 070 (2015).

[12] T. Aoki, M. Hirasawa, Y. Ito, J. Nishimura and A. Tsuchiya, On the structure of the emergent 3d expanding space in the Lorentzian type IIB matrix model, *PTEP* **2019**, p. 093B03 (2019).

[13] J. Nishimura and A. Tsuchiya, Complex Langevin analysis of the spacetime structure in the Lorentzian type IIB matrix model, *JHEP* **06**, p. 077 (2019).

[14] K. Hatakeyama, K. Anagnostopoulos, T. Azuma, M. Hirasawa, Y. Ito, J. Nishimura, S. Papadoudis and A. Tsuchiya, Relationship between the Euclidean and Lorentzian versions of the type IIB matrix model, in *38th International Symposium on Lattice Field Theory*, arXiv:2112.15368 [hep-lat]

[15] M. Hirasawa, K. Anagnostopoulos, T. Azuma, K. Hatakeyama, Y. Ito, J. Nishimura, S. Papadoudis and A. Tsuchiya, A new phase in the Lorentzian type IIB matrix model and the emergence of continuous

18

space-time, in *38th International Symposium on Lattice Field Theory*, `arXiv:2112.15390 [hep-lat]`

[16] G. Parisi, ON COMPLEX PROBABILITIES, *Phys. Lett. B* **131**, 393 (1983).

[17] J. R. Klauder, Coherent State Langevin Equations for Canonical Quantum Systems With Applications to the Quantized Hall Effect, *Phys. Rev. A* **29**, 2036 (1984).

[18] G. Aarts, F. A. James, E. Seiler and I.-O. Stamatescu, Adaptive stepsize and instabilities in complex Langevin dynamics, *Phys. Lett. B* **687**, 154 (2010).

[19] G. Aarts, E. Seiler and I.-O. Stamatescu, The Complex Langevin method: When can it be trusted?, *Phys. Rev. D* **81**, p. 054508 (2010).

[20] G. Aarts, F. A. James, E. Seiler and I.-O. Stamatescu, Complex Langevin: Etiology and Diagnostics of its Main Problem, *Eur. Phys. J. C* **71**, p. 1756 (2011).

[21] J. Nishimura and S. Shimasaki, New Insights into the Problem with a Singular Drift Term in the Complex Langevin Method, *Phys. Rev. D* **92**, p. 011501 (2015).

[22] K. Nagata, J. Nishimura and S. Shimasaki, Justification of the complex Langevin method with the gauge cooling procedure, *PTEP* **2016**, p. 013B01 (2016).

[23] K. Nagata, J. Nishimura and S. Shimasaki, Argument for justification of the complex Langevin method and the condition for correct convergence, *Phys. Rev. D* **94**, p. 114515 (2016).

[24] Y. Ito and J. Nishimura, The complex Langevin analysis of spontaneous symmetry breaking induced by complex fermion determinant, *JHEP* **12**, p. 009 (2016).

[25] S. Brahma, R. Brandenberger and S. Laliberte, Emergent Cosmology from Matrix Theory, `arXiv:2107.11512 [hep-th]` .

[26] H. C. Steinacker, Gravity as a Quantum Effect on Quantum Space-Time, `arXiv:2110.03936 [hep-th]` .

[27] F. R. Klinkhamer, Towards a numerical solution of the bosonic masterfield equation of the IIB matrix model, `arXiv:2110.15309 [hep-th]` .

[28] K. Hatakeyama, A. Matsumoto, J. Nishimura, A. Tsuchiya and A. Yosprakob, The emergence of expanding space–time and intersecting D-branes from classical solutions in the Lorentzian type IIB matrix model, *PTEP* **2020**, p. 043B10 (2020).

Digital Quantum Simulation of the Schwinger model with Topological term

Masazumi Honda

Yukawa Institute for Theoretical Physics, Kyoto University,
Sakyo-ku, Kyoto 606-8502, Japan
E-mail: masazumi.honda@yukawa.kyoto-u.ac.jp

Recently computational resource of quantum computers sounds growing well. In this article, we discuss how we can apply this development to numerically simulate quantum field theories. In contrast to the conventional approach by (Marlov chain) Monte Carlo method suffering from the infamous sign problem, we work in Hamilton formalism and adopt quantum algorithms which do not rely on Monte Carlo sampling. After brief discussion on how to put quantum field theories on quantum computers, we present our recent numerical results on the charge-q Schwinger model, where q is an electric charge of a Dirac fermion. We observe an exotic phenomena such as negative string tension behavior in potential between heavy charged particles which essentially come from presense of non-small θ-angle.

Keywords: Quantum computation, Gauge theory, Schwinger model, Lattice gauge theory, Confinement

1. Introduction

It seems that resource of quantum computers has been recently growing well. In particular, readers would remember the news in 2019 that Google claimed to achieve quantum supremacy. While the news gave contraversy on whether or not it is really quantum supremacy, today quantum computer is one of hot science topics in public media. It is now possible for anyone to use (small-scale) quantum computer in the cloud for free. Although public news on quantum computers usually mention industrial applications such as sequrity, new medicine and so on, in this article, we would like to consider how these developments can help us to understand physical systems. In particular, we discuss applications of quantum computation to numerical simulations of quantum field theories (QFT), and present some of our recent results in this context [1–3]. Since QFT is a common language in various fields such as high energy physics, nuclear physics, cosmology and condensed matters etc., it is expected to induce developments in various fields.

2. Conventional numerical approach to quantum field theory

For many purposes, one of the main motivations to use a quantum computer would be to perform fast numerical calculations. The purpose of this article, the application of quantum field theory to numerical simulations, is not an exception but in this case the motivation is more specific. To explain this, we discuss how numerical simulations of quantum field theoriesa have been usually performed. In the conventional approach, we use Lagrangian (path integral) formalism and the expectation value of the operator $\mathcal{O}$ for Euclidean case is written as

$$\langle \mathcal{O}(\Phi) \rangle = \frac{\int D\Phi \; \mathcal{O}(\Phi) e^{-S[\Phi]}}{\int D\Phi \; e^{-S[\Phi]}}. \tag{1}$$

where $S[\Phi]$ is the action and "integral domain" is over all possible values of the field Φ at each point in the spacetime. Of course we cannot directly perform a numerical evaluation of the infinite dimensional integral and we need some regularization. The most standard way to do this is to cut the spacetime into a lattice with finite size (lattice regularization). To reproduce to the original theory, we need to take the continuous limit where the lattice spacing goes to zeroa. Numerical evaluation of the regularized integral is usually done by an algorithm called the (Markov chain) Monte Carlo method, where we regard the Boltzmann weight ($\propto e^{-S}[\Phi]$) as the probability that realizes the field configurationΦ. Then we approximate the integral by using the average over the generated samples:

$$\langle \mathcal{O}(\Phi) \rangle \simeq \frac{1}{\sharp(\text{samples})} \sum_{j \in \text{samples}} \mathcal{O}(\Phi_j). \tag{2}$$

So far, the conventional approch has been successful to some extent. One of the significant successes is the derivation of the nuclear force from the first principle by lattice QCD (quantum chromodynamics)[4]. However, if the Boltzmann weights are not positive real numbers, the probability interpretation cannot be applied directly and we need some tricks[b]. In particular,

---

[a]If the original spacetime has infinite volume, then we should also take a infinite volume limit.

[b]There are various efforts within the framework of the path integral formalism, which have limited success as sign problem becomes stronger (see e.g. the paper[5]).

when the integrand is highly oscillating, it is likely hard to do efficient sampling. This infamous sign problem often occurs physically, for example, in the presence of topological terms, chemical potentials and real time. All these situations are closely related to very important physical problems.

What would happen if we work in the Hamiltonian formalism rather than the Lagrangian formalism? In the Hamiltonian formalism, the sign problem does not exist from the beginning since the problem to be solved is not a (path) integral. However, there is a reason why the Hamiltonian formalism has not been used much in numerical simulations of quantum field theory: since state space of QFT is typically infinite, we need to regularize it but the dimension of the state space after regularization typically increases exponentially with the number of "degrees of freedom". In other words, naively, computers have to memorize exponentially large vectors corresponding to the states and multiply huge matrices corresponding to operators. This is the case for classical computer. What about quantum computers? For quantum computers, at least for some problems in quantum field theory, there are known algorithms with exponential improvement in computational complexity compared to classical computation. Here we consider what will be possible when the resources of quantum computers are expected to be sufficiently developed in the future.

3. Quantum field theory as qubits

Let us see how we can put quantum field theories on quantum computers. In gate-type quantum computer, the basic unit is quantum bit (qubit) which is a quantum system with two dimensional Hilbert space. In order to directly apply quantum algorithms to QFT, it is convenient to rewrite QFT as a spin system. Here we discuss how we can achieve this in terms of simple examples.

3.1. *Fermion field*

It is easiest for fermions because state space associated with fermions becomes finite dimensional by simply cutting the spatial directions by a finite lattice. This is essentially because of the Pauli's exclusion principle. Let us explicitly see this in a free Dirac fermion in $1 + 1$ dimensions. We cut the one dimensional space by a lattice with N sites and lattice spacing a. If we choose staggered fermion[6] as a lattice fermion, the Hamiltonian of the lattice theory is given by

$$H = -\frac{i}{2a}\sum_{n=0}^{N-2}\left[\chi_n^\dagger\chi_{n+1} - \chi_{n+1}^\dagger\chi_n\right] + m\sum_{n=0}^{N-1}(-1)^n\chi_n^\dagger\chi_n, \quad (3)$$

where m is fermion mass and the lattice fermion operators $(\chi_n, \chi_n^\dagger)$ satisfy

$$\{\chi_m, \chi_n^\dagger\} = \delta_{mn}. \quad (4)$$

In this case the dimension of the state space is finite in dimension 2^N.

To rewrite the system in a form that we can directly apply quantum algorithms, we map it to a spin system. To do this, we need a spin operator that satisfies the canonical anti-commutation relation (4). While such a spin operator is not unique, the most traditional one is the so-called Jordan-Wigner transformation[7]:

$$\chi_n = \frac{X_n - iY_n}{2}\prod_{j=1}^{n-1}(-iZ_j), \quad (5)$$

where (X_n, Y_n, Z_n) is the Pauli matrix $(\sigma_1, \sigma_2, \sigma_3)$ located at site n. Then we find

$$H = \frac{1}{4a}\sum_{n=0}^{N-2}\left[X_nX_{n+1} + Y_nY_{n+1}\right] + \frac{m}{2}\sum_{n=0}^{N-1}(-1)^nZ_n. \quad (6)$$

This is the same as the Hamiltonian of the XY-model with (space dependent) longitudinal magnetic field.

3.2. Scalar field

Next let us consider scalar field. The large difference from the fermionic case is that the state space is still infinite dimensional even on lattice and further regularization is required. This is because lattice scalar field theory is technically equivalent to multi-particle quantum mechanics with appropriate interactions which have infinite dimensional Hlibert space. To explain how to put lattice scalar field theory on quantum computers, let us consider a single particle quantum mechanics for simplicity:

$$H(x, p) = \frac{1}{2}p^2 + \frac{\omega^2}{2}x^2 + V(x), \quad (7)$$

where $V(x)$ is a potential and

$$[x, p] = i. \quad (8)$$

In principle there are many ways to truncate the Hilbert space but here we use the most naive truncation by harmonic oscillator basis[c]. To do this, let us introduce annhilation and creation operators for the harmonic oscillator:

$$x = \frac{1}{\sqrt{2\omega}}(a + a^\dagger), \quad p = i\sqrt{\frac{\omega}{2}}(a - a^\dagger). \tag{9}$$

The annihilation operator in the number operator basis n is

$$a = \sum_{n=0}^{\infty} \sqrt{n+1}|n\rangle\langle n+1|. \tag{10}$$

Let us truncate the Hilbert space such that it is spanned by $|0\rangle, |1\rangle, \cdots, |\Lambda-1\rangle$ and replace the annihilation operator by

$$a \quad \to \quad a_\Lambda := \sum_{n=0}^{\Lambda-2} \sqrt{n+1}|n\rangle\langle n+1|. \tag{11}$$

Then we consider the Hamiltonian

$$H(x, p) \quad \to \quad H_\Lambda := H(x_\Lambda, p_\Lambda), \tag{12}$$

where

$$x_\Lambda = \frac{1}{\sqrt{2\omega}}(a_\Lambda + a_\Lambda^\dagger), \quad p_\Lambda = i\sqrt{\frac{\omega}{2}}(a_\Lambda - a_\Lambda^\dagger). \tag{13}$$

We can map the truncated system to a spin system as follows. First we consider a binary representation of n:

$$n = b_{K-1}2^{K-1} + b_{K-2}2^{K-2} + \cdots + b_0 2^0 \quad (K = \log \Lambda), \tag{14}$$

and express the basis $|n\rangle$ as

$$|n\rangle = |b_0\rangle \cdots |b_K\rangle. \tag{15}$$

Then each component of the annihilation operator is given by

$$|n\rangle\langle n+1| = \otimes_{\ell=0}^{K-1} \left(|b_\ell'\rangle\langle b_\ell|\right), \tag{16}$$

which can be mapped to a spin system via

$$|0\rangle\langle 0| = \frac{1-Z}{2}, \quad |1\rangle\langle 1| = \frac{1+Z}{2},$$
$$|0\rangle\langle 1| = \frac{X+iY}{2}, \quad |1\rangle\langle 0| = \frac{X-iY}{2}. \tag{17}$$

This is a qubit description of the single particle quantum mechanics. It is straightforward to generalize this method to multi-particle case.

---

[c]See e.g. the papers[8-10] for other basis.

3.3. Gauge field

The case of gauge fields is more complicated and depends on spacetime dimensions and and boundary conditions. This is realted to the fact that in gauge theory, there are not only physical states but also unphysical states, and whether or not the dimension of the physical state space is infinite depends on situations. To see this concretely, let us consider a $1+1$ dimensional quantum electrodynamics coupled to charge-q Dirac fermion. We will call this theory the charge-q Schwinger model or simply the Schwinger model[11,12]. When the space is cut into a lattice, the Hamiltonian of this theory is given by[13]

$$
H = -\frac{i}{2a} \sum_{n=0}^{N-2} \left[\chi_n^\dagger e^{iq\phi_n} \chi_{n+1} - \chi_{n+1}^\dagger e^{-iq\phi_n} \chi_n \right]
$$
$$
+ m \sum_{n=0}^{N-1} (-1)^n \chi_n^\dagger \chi_n + \frac{g^2 a}{2} \sum_{n=0}^{N-2} \left(L_n + \frac{\theta}{2\pi} \right)^2, \tag{18}
$$

where g is gauge coupling and θ is theta angle. The lattice gauge field operators (ϕ_n, L_n) satisfy the canonical commutation relation

$$
[\phi_m, L_n] = i\delta_{mn}. \tag{19}
$$

Now the state space associated with the gauge field is infinite dimensional since ϕ_n is bosonic. However, this is the state space including unphysical states and the physical states are restricted by the Gauss law:

$$
L_n - L_{n-1} = q \left(\chi_n^\dagger \chi_n - \frac{1 - (-1)^n}{2} \right). \tag{20}
$$

This relates L_n to its neighbours and the fermionic operator. When we take the open boundary condition, we can rewrite L_n purely in terms of the fermionic operator and ϕ_n can also be absorbed by a gauge transformation. Therefore the Schwinger model with the Gauss law is rewritten as a fermionic system and we can further map it to a spin system by e.g. the Jordan-Wigner transformation. Thus we can directly apply the quantum algorithm to the Schwinger model.

4. Quantum algorithm to prepare vacuum

Here we would like to construct the ground state by a quantum algorithm. While there are several known algorithms, here we use an algorithm called adiabatic state preparation. Suppose that we would like to constract the ground state of the Hamiltonian H_{target} of the target system. First, we prepare an initial Hamiltonian H_0 whose ground state $|GS_0\rangle$ is known and non-degenerate. Therefore H_0 is practically taken to be a Hamiltonian of a simple system. Next, we introduce a time-dependent Hamiltonian $H_A(t)$ satisfying

$$H_A(0) = H_0, \quad H_A(T) = H_{\text{target}}, \tag{21}$$

and we take $H_A(t)$ to change more slowly for larger T. Then we use the adiabatic theorem. That is, if the ground state of $H_A(t)$ has non-degenerate for any t, then the ground state $|GS\rangle$ of H_{target} is obtained by the following time evolution:

$$|GS\rangle = \lim_{T\to\infty} \mathcal{T}\exp\left(-i\int_0^T dt H_A(t)\right)|GS_0\rangle. \tag{22}$$

In actual simulations, the time evolution operator is approximated by taking finite T and discretizing the integral in the exponent. This gives errors due to the fact that the constructed state in pratical simulation is not exactly the ground state. Using the approximate ground state constructed in this way, we can calculate the expectation value under the ground state approximately.

5. Recent simulation results

Using the adiabatic approximation, we can construct the ground state of the Schwinger model and calculate various physical quantities. Here we present a part of numerical results of our simulations[1–3] (see e.g. the papers[14–18] for other simulations of the Schwinger model). The Schwinger model has been analyzed by various methods. Since it is $1+1$ dimensional, a powerful anaytic approach by bosonization is available. In particular, it is known that we can exaclty solve the massless case ($m = 0$)[19–22]. When the fermion mass m is nonzero, there is no known exact solution but we can get a good approximation by mass perturbation theory[23,24] for small m. Regarding the conventional Monte Carlo approach, it is known to be difficult when θ is not small due to the sign problem while small θ region is accessible.

There is one thing before presenting the results: real quantum computer has errors due to interactions with einvironment. Therefore we need to take error correction into account in order to obtain error-free results. However, it is known that this requires huge computational resources and this is a major obstacle in development on the technology side. To test quantum algorithms and estimate the computational resources required, people often use a tool called a (classical) simulator. A simulator is a tool to simulate

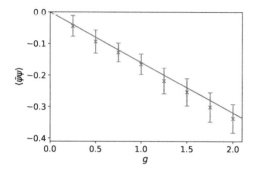

Fig. 1. Chiral condensate in the $q = 1$ Schwinger model after thermodynamic and continuum limits against the gauge coupling g for the massless case ($m = 0$). The red straight line represents the exact result.

a quantum computer by a classical computer and we can use almost the same code to run a real quantum computer. Here we use a simulator called "qasm simulator" provided by IBM.

5.1. The $q = 1$ Schwinger model without probe charges

First, we focus on the $q = 1$ case, which has been investigated well. Here we consider a quantity called the chiral condensate[1]. In the continuum theory, this quantity is defined as the expectation value of the fermion mass operator $\bar{\psi}\psi(x)$ under the groundstate:

$$\langle \text{GS} | \bar{\psi}\psi(x) | \text{GS} \rangle. \tag{23}$$

We compute the chiral condensate for various values of the parameters (a, N), and take the infinite volume limit $N \to \infty$ and the continuous limit $a \to 0$. In fig. 1, we plot the chiral condensate for the massless case against g and find agreement with the exact result. In Fig. 2, we plot the chiral condensate[d] at $g = 1$ against m for $\theta = 0$ and $\theta = 3\pi/5$. When m is small, the simulation and mass perturbation results agree, whereas when m is large, they deviate from each other. This is intepreted as violation of mass petruatation theory for non-small m. In the context of quantum simulations of field theory, this is the first result in which the continuous limit is taken seriously.

---

[d]We regularize it by subtracting the free result.

27

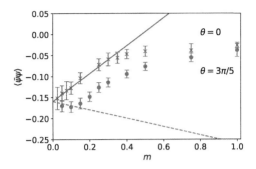

Fig. 2. The chiral condensate in the $q = 1$ Schwinger model for massive case at $g = 1$ for $\theta = 0$ (blue ×'s) and $\theta = 3\pi/5$ (green circles). The lines denote the result obtained by the mass perturbation theory.

5.2. With probe charges

Next let us consider the $q \neq 1$ case[2,3]. We also put two probe charges $\pm q_p$ and consider the potential between them. According to the mass perturbation theory[25], the potential changes its qualitative bahavior as changing the parameters:

$$V(\ell) = \sigma \ell \quad \text{for } g\ell \gg 1, \tag{24}$$

where ℓ is the distance between the probes and σ is the string tension

$$\sigma = -m\frac{e^{\gamma}qg}{2\pi^{3/2}}\left[\cos\frac{\theta + 2\pi q_p}{q} - \cos\frac{\theta}{q}\right] + \mathcal{O}(m^2). \tag{25}$$

At first sight, this seems to show a confinement behavior but note that the tension σ is not always positive. In particular, we find $\sigma = 0$ when q_p is an integer multiple of q. This implies screening of the charges. We can also see that σ can be sometimes negative when q_p/q is not an integer. The reason why this can happen is closely related to the generalized global symmetry, but we will not go into details here (see e.g. the papers[3,26,27]).

To calculate the potential, we measure the difference of the ground state energies in the theories with and without the probe charges. In the $(1 + 1)$-dimensional $U(1)$ gauge theory, the effect of the probe charges is taken into account by replacing θ by a space-dependent θ_n such that its value suddenly changes by $2\pi q_p$. In this way we can compute the potential. Figure 3 shows the potential for $q = 3$ and $q_p = -1$ for various values of θ. The simulation results show a linear behavior for all values of θ. while

28

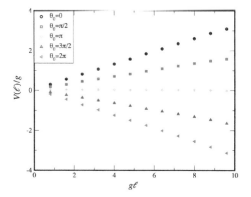

Fig. 3. The potential $V(\ell)/g$ between the probe charges $(q_p = -1)$ is plotted in the $q = 3$ Schwinger model for various values of $\theta = \theta_0$. Here we take $m = 0.15$, $N = 25$ and $ga = 0.40$.

signs of its slopes change as changing θ at the same timing as the result of the mass perturbation theory.

6. Discussion

In this article, we discussed applications of quantum computation to numerical simulations of quantum field theories. This topic is still in its infancy and there are still many things to explore. There would be future developments both in hardware and software aspects. While we presented recent numerical results obtained by a classical simulator, we need use real quantum computers to have speed up of quantum algorithms.

Regarding the hardware perspective, although an ideal quantum computer is a large machine with error correction, machines available in near future are NISQ (Noisy Intermediate-Scale Quantum device)[28], where errors are not negligible. For such a machine, it is desirable to apply an algorithm that uses as few quantum gates as possible. While the adiabatic state preparation used in this article is guaranteed to be correct under the assumptions of the adiabatic theorem, it is not suitable for NISQ devise because it uses many quantum gates to approximate the time evolution. The algorithm expected to be suitable for NISQ is the one that uses only a small number of gates such as classical-quantum hybrid algorithm. An algorithm to construct the ground state along such a direction is based on the variational method.

Regarding the software perspective, we have not yet established an efficient way to put most field theories on a quantum computer. In particular, in other gauge theories above $2+1$ dimensions such as quantum chromodynamics, it is hard to establish to put theories on quantum computer while respecting gauge symmetry. So far, a simple lattice regularization has been applied but this may not be best for some theories[30]. In addition, there are many interesting states other than the ground state in QFT, but this aspect has not been explored well yet in the context of quantum computation. One of the most interesting states in high energy physics are ones treated in the scattering problem, in which multiple particles are moving with a fixed momentum. If we can construct such a state, then we would be able to understand what is "actually" happening in the accelerator experiments. In this article, we have focused on QFT, but it is also interesting to consider applications to quantum gravity such as string theory[29]. I hope that we will be able to literally see time evolution of our universe on quantum computers in the future.

Acknowledgments

The author would like to thank Alexander Buser, Bipasha Chakraborty, Hrant Gharibyan, Masanori Hanada, Etsuko Itou, Taku Izubuchi, Yuta Kikuchi, Junyu Liu, Lento Nagano, Takuya Okuda, Yuya Tanizaki and Akio Tomiya for collaborations on applications of quantum computation to high energy physics[1-3,29,30]. He is supported by MEXT Q-LEAP, JST PRESTO Grant Number JPMJPR2117, Japan and JSPS Grant-in-Aid for Transformative Research Areas (A) JP21H05190.

References

1. B. Chakraborty, M. Honda, T. Izubuchi, Y. Kikuchi and A. Tomiya, Digital Quantum Simulation of the Schwinger Model with Topological Term via Adiabatic State Preparation (1 2020).
2. M. Honda, E. Itou, Y. Kikuchi, L. Nagano and T. Okuda, Classically emulated digital quantum simulation for screening and confinement in the Schwinger model with a topological term, *Phys. Rev. D* **105**, p. 014504 (2022).
3. M. Honda, E. Itou, Y. Kikuchi and Y. Tanizaki, Negative string tension of higher-charge Schwinger model via digital quantum simulation (10 2021).

4. N. Ishii, S. Aoki and T. Hatsuda, The Nuclear Force from Lattice QCD, *Phys. Rev. Lett.* **99**, p. 022001 (2007).
5. G. Aarts, Introductory lectures on lattice QCD at nonzero baryon number, *J. Phys. Conf. Ser.* **706**, p. 022004 (2016).
6. L. Susskind, Lattice Fermions, *Phys. Rev.* **D16**, 3031 (1977).
7. P. Jordan and E. Wigner, Über das paulische äquivalenzverbot, *Zeitschrift für Physik* **47**, 631 (Sep 1928).
8. S. P. Jordan, K. S. M. Lee and J. Preskill, Quantum Algorithms for Quantum Field Theories, *Science* **336**, 1130 (2012).
9. S. P. Jordan, K. S. M. Lee and J. Preskill, Quantum Computation of Scattering in Scalar Quantum Field Theories (2011), [Quant. Inf. Comput.14,1014(2014)].
10. N. Klco and M. J. Savage, Digitization of scalar fields for quantum computing, *Phys. Rev.* **A99**, p. 052335 (2019).
11. J. S. Schwinger, Gauge Invariance and Mass, *Phys. Rev.* **125**, 397 (1962).
12. S. R. Coleman, R. Jackiw and L. Susskind, Charge Shielding and Quark Confinement in the Massive Schwinger Model, *Annals Phys.* **93**, p. 267 (1975).
13. J. B. Kogut and L. Susskind, Hamiltonian Formulation of Wilson's Lattice Gauge Theories, *Phys. Rev.* **D11**, 395 (1975).
14. E. A. Martinez *et al.*, Real-time dynamics of lattice gauge theories with a few-qubit quantum computer, *Nature* **534**, 516 (2016).
15. C. Muschik, M. Heyl, E. Martinez, T. Monz, P. Schindler, B. Vogell, M. Dalmonte, P. Hauke, R. Blatt and P. Zoller, U(1) Wilson lattice gauge theories in digital quantum simulators, *New J. Phys.* **19**, p. 103020 (2017).
16. N. Klco, E. F. Dumitrescu, A. J. McCaskey, T. D. Morris, R. C. Pooser, M. Sanz, E. Solano, P. Lougovski and M. J. Savage, Quantum-classical computation of Schwinger model dynamics using quantum computers, *Phys. Rev.* **A98**, p. 032331 (2018).
17. C. Kokail *et al.*, Self-verifying variational quantum simulation of lattice models, *Nature* **569**, 355 (2019).
18. G. Magnifico, M. Dalmonte, P. Facchi, S. Pascazio, F. V. Pepe and E. Ercolessi, Real Time Dynamics and Confinement in the $\mathbb{Z}_n$ Schwinger-Weyl lattice model for 1+1 QED (2019).
19. J. E. Hetrick and Y. Hosotani, QED ON A CIRCLE, *Phys. Rev.* **D 38**, p. 2621 (1988).

20. J. E. Hetrick, Y. Hosotani and S. Iso, The Massive multi - flavor Schwinger model, *Phys. Lett.* **B350**, 92 (1995).

21. Y. Hosotani, R. Rodriguez, J. E. Hetrick and S. Iso, Confinement and chiral dynamics in the multiflavor Schwinger model, 382 (1996).

22. Y. Hosotani, Gauge theory model: Quark dynamics and antiferromagnets, 445 (1996).

23. C. Adam, Normalization of the chiral condensate in the massive Schwinger model, *Phys. Lett.* **B440**, 117 (1998).

24. C. Adam, Massive Schwinger model within mass perturbation theory, *Annals Phys.* **259**, 1 (1997).

25. D. J. Gross, I. R. Klebanov, A. V. Matytsin and A. V. Smilga, Screening versus confinement in (1+1)-dimensions, *Nucl. Phys. B* **461**, 109 (1996).

26. T. Misumi, Y. Tanizaki and M. Ünsal, Fractional θ angle, 't Hooft anomaly, and quantum instantons in charge-q multi-flavor Schwinger model, *JHEP* **07**, p. 018 (2019).

27. Z. Komargodski, K. Ohmori, K. Roumpedakis and S. Seifnashri, Symmetries and strings of adjoint QCD_2, *JHEP* **03**, p. 103 (2021).

28. J. Preskill, Quantum Computing in the NISQ era and beyond, *Quantum* **2**, p. 79 (August 2018).

29. H. Gharibyan, M. Hanada, M. Honda and J. Liu, Toward simulating superstring/M-theory on a quantum computer, *JHEP* **07**, p. 140 (2021).

30. A. J. Buser, H. Gharibyan, M. Hanada, M. Honda and J. Liu, Quantum simulation of gauge theory via orbifold lattice, *JHEP* **09**, p. 034 (2021).

Multi-Soliton scattering of the Anti-Self-Dual Yang-Mills Equations in 4-dimensional split signature

Shan-Chi Huang*

Graduate School of Mathematics, Nagoya University,
Nagoya, 464-8602, Japan
* E-mail: x18003x@math.nagoya-u.ac.jp

We construct the ASDYM 1-solitons and multi-solitons for split signature and interpret them as soliton walls. We show that the gauge group is $G = SU(2)$ for the entire intersecting soliton walls, and $SU(3)$ for each soliton walls in the asymptotic region. This is joint research partially with Masashi Hamanaka, Claire R. Gilson, and Jonathan Nimmo.

Keywords: Anti-self-dual Yang-Mills equation; Soliton scattering; Exact solvable model; N=2 Strings.

1. Introduction

The anti-self-dual Yang-Mills (ASDYM) system is notable as an exactly solvable model and play an indispensable role in many different fields of theoretical and mathematical physics, such as quantum field theory, twistor theory[8], and integrable systems[8]. Especially the ASDYM equations reveal the essence of the lower-dimensional integrable systems because almost all of the soliton equations are the dimensionally reduced equations of ASDYM according to the Ward conjecture[12]. Among these lower-dimensional soliton equations, dimensionally reduced from the 4-dimensional split signature $(+, +, -, -)$ (or called the Ultrahyperbolic[8] spacetime) are the majority. On the other hand, the ASDYM equations in the 4-dimensional split signature is the EOM of effective action for $N = 2$ open string theories[11] and seem to have good compatibility with the lower-dimensional integrable systems. Although most physicists didn't pay too much attention on $N = 2$ strings and there has been little progress in this direction for a long time, we still believe that the connections between the lower-dimensional integrable

*This work is supported by Grant-in-Aid for Scientific Research (18K03274) and the scholarship of Japan-Taiwan Exchange Association.

systems and N = 2 open string theories are worth studying and not just for seeking the physical reality.

In (1+1)-dimensional integrable systems, the KdV equation $-4u_t + 6uu_x + u_{xxx} = 0$ might be the most notable one among the exactly solvable models and the typical KdV 1-soliton is in the form of

$$u(x,t) := 2\frac{\partial^2}{\partial x^2}(\log \cosh X) = 2\kappa^2 \text{sech}^2 X, \quad X := \kappa x + \kappa^3 t + \delta \qquad (1)$$

which is a travelling wave solution with a constant velocity $-\kappa^2$ and a constant amplitude $2\kappa^2$, and hence it preserves its shape over time. In addition, $u(x,t)$ is an even function symmetric with respect to $X = 0$ and $u, u_x, u_{xx}, u_{xxx} \to 0$ as $x \to \pm\infty$. Therefore, the distribution of $u(x,t)$ is localized within a region centered on $X = 0$. On the other hand, the KdV equation is also known for having infinitely conserved quantities[9], more specifically, the mass $\int u dx$, the momentum $\int u^2 dx$, the energy $\int [2u^3 - (u_x)^2] dx$, ..., and so on. In particular, the energy density

$$2u^3 - (u_x)^2 = 16\kappa^6 \left(2\text{sech}^6 X - \text{sech}^4 X\right), \quad X := \kappa x + \kappa^3 t + \delta \qquad (2)$$

possesses the behavior of 1-soliton as well. Inspired by this, we simply consider the pure Yang-Mills action density $\text{Tr}F_{\mu\nu}F^{\mu\nu}$ as an analogue of energy density, and try to figure out whether the 1-solitonic behavior like (2) exists for the ASDYM equations in 4-dimensions or not. Fortunately, the answer is yes and our result[4] is

$$\text{Tr}F_{\mu\nu}F^{\mu\nu} \propto \left(2\text{sech}^2 X - 3\text{sech}^4 X\right), \qquad (3)$$

where X is a nonhomogeneous linear function of real coordinates x^1, x^2, x^3, x^4. Especially, the gauge group is G = SU(2)[4] for the split signature and hence it means our ASDYM 1-soliton could be some candidate of physical object in N = 2 open string theories.

One more well-known fact is that the KdV multi-soliton

$$u(x,t) := 2\frac{\partial^2}{\partial x^2}(\log \tau_n) \qquad (4)$$

can be represented elegantly as the Wronskian determinant

$$\tau_n := \text{Wr}(f_1, f_2, ..., f_n) := \begin{vmatrix} f_1^{(0)} & f_2^{(0)} & \cdots & f_n^{(0)} \\ f_1^{(1)} & f_2^{(1)} & \cdots & f_n^{(1)} \\ \vdots & \vdots & \ddots & \vdots \\ f_1^{(n-1)} & f_2^{(n-1)} & \cdots & f_n^{(n-1)} \end{vmatrix}, \qquad (5)$$

where $f_i^{(m)} := \dfrac{\partial^m f_i}{\partial x^m}$, $f_i := \cosh X_i$, $X_i := \kappa_i x + \kappa_i^3 t + \delta_i$. $\qquad$ (6)

Now a natural question to ask is whether the Wronskian type multi-solitons exist for the ASDYM equations or not. If the answer is yes and the gauge group can be verified to be unitary, then what is interpretation for such ASDYM multi-solitons in N = 2 string theories.

2. J-matrix formulation of ASDYM and Darboux transformation

Before going ahead to the main topic, we need to introduce some prior knowledge. Firstly, the ASDYM equations on various real spaces can be unified into 4-dimensional complex flat spacetime with the metric $ds^2 = 2(dzd\widetilde{z} - dwd\widetilde{w})$ and we can easily get the split signature $(+, +, -, -)$ by imposing some conditions on the complex coordinates, for instance

$$\begin{pmatrix} z & w \\ \widetilde{w} & \widetilde{z} \end{pmatrix} = \frac{1}{\sqrt{2}} \begin{pmatrix} x^1 + x^3 & x^2 + x^4 \\ -(x^2 - x^4) & x^1 - x^3 \end{pmatrix}, \ x^1, x^2, x^3, x^4 \in \mathbb{R}. \tag{7}$$

Here we set the gauge group to be $G = GL(N, \mathbb{C})$ in general. The complex representation of ASDYM equations are

$$F_{zw} = 0, \ F_{\widetilde{z}\widetilde{w}} = 0, \ F_{z\widetilde{z}} - F_{w\widetilde{w}} = 0 \tag{8}$$

which can be cast in a gauge independent formulation, called the Yang equation[1,13]

$$\partial_{\widetilde{z}}[(\partial_z J)J^{-1}] - \partial_{\widetilde{z}}[(\partial_z J)J^{-1}] = 0, \tag{9}$$

where J is an $N \times N$ matrix called Yang's J-matrix. An advantage of this formulation is that the anti-self-dual (ASD) gauge fields can be reformulated by the decomposition of $J = \widetilde{h}^{-1}h$ as

$$A_z = -(\partial_z h)h^{-1}, A_w = -(\partial_w h)h^{-1}, A_{\widetilde{z}} = -(\partial_{\widetilde{z}}\widetilde{h})\widetilde{h}^{-1}, A_{\widetilde{w}} = -(\partial_{\widetilde{w}}\widetilde{h})\widetilde{h}^{-1}. \tag{10}$$

Sometimes we use a convenient gauge $\widetilde{h} = 1$ to simplify the gauge fields as

$$A_z = -(\partial_z J)J^{-1}, \ A_w = -(\partial_w J)J^{-1}, \ A_{\widetilde{z}} = 0, \ A_{\widetilde{w}} = 0. \tag{11}$$

Now we introduce a novel Lax representation[10] of ASDYM equations :

$$\begin{cases} L(\phi) := [\partial_w - (\partial_w J)J^{-1}]\phi - (\partial_{\widetilde{z}}\phi)\zeta = 0 \\ M(\phi) := [\partial_z - (\partial_z J)J^{-1}]\phi - (\partial_{\widetilde{w}}\phi)\zeta = 0 \end{cases}, \ (G = GL(N, \mathbb{C})) \tag{12}$$

where the spectral parameter ζ is any right-action $N \times N$ constant matrix rather than scalar. Under the compatible condition $L(M(\phi)) - M(L(\phi)) = 0$, the ASDYM equations (Yang equation) can be derived from the linear system (12). Furthermore, if we define a Darboux Transformation[10] as

$$\widetilde{\phi} = \phi\zeta - \psi\Lambda\psi^{-1}\phi, \ \ \widetilde{J} = -\psi\Lambda\psi^{-1}J, \tag{13}$$

where $\phi(\zeta)$ denotes the general solution of (12) and we use a new notation $\psi(\Lambda)$ to denote a specified solution of (12). Then the linear system (12) is form invariant under the Darboux transformation (13). In other words,

$$\begin{cases} \widetilde{L}(\widetilde{\phi}) := [\partial_w - (\partial_w \widetilde{J})\widetilde{J}^{-1}]\widetilde{\phi} - (\partial_{\bar{z}}\widetilde{\phi})\zeta = 0 \\ \widetilde{M}(\widetilde{\phi}) := [\partial_z - (\partial_z \widetilde{J})\widetilde{J}^{-1}]\widetilde{\phi} - (\partial_{\widetilde{w}}\widetilde{\phi})\zeta = 0 \end{cases} . \tag{14}$$

Now we can choose a seed solution $J = J_1$ of the Yang equation (9) and substitute it into (12) to solve a specified solution $\psi_1(\Lambda_1)$. After 1-iteration of the Darboux transformation (13), we get a new solution $\widetilde{J} = J_2$. By repeating the same process[6], we can obtain a series of J-matrices.

$$\overbrace{J_1}^{\text{Seed solution}} \xrightarrow{\text{Dar}} J_2 \xrightarrow{\text{Dar}} J_3 \xrightarrow{\text{Dar}} J_4 \xrightarrow{\text{Dar}} \dots \xrightarrow{\text{Dar}} J_{n+1} \xrightarrow{\text{Dar}} \dots$$

and the J-matrix J_{n+1} can be written concisely in terms of the Wronskian[3] type quasideterminant (Cf: (5)) :

$$J_{n+1} = \begin{vmatrix} \psi_1 & \psi_2 & \cdots & \psi_n & 1 \\ \psi_1\Lambda_1 & \psi_2\Lambda_2 & \cdots & \psi_n\Lambda_n & 0 \\ \vdots & \vdots & \ddots & \vdots & \vdots \\ \psi_1\Lambda_1^n & \psi_2\Lambda_2^n & \cdots & \psi_n\Lambda_n^n & \boxed{0} \end{vmatrix} J_1, \tag{15}$$

where $\psi_i(\Lambda_i)$ denote n specified solutions of (12) with matrix size $N \times N$. We use the term quasi-Wronskian to call it for short. In fact, since the elements of J_i are noncommutative, the quasiderminant can be considered roughly as a noncommutative version of determinant. By the definition[2] , we can decompose (15) into 4 blocks as

$$\begin{vmatrix} A_{nN \times nN} & B_{nN \times N} \\ C_{N \times nN} & \boxed{D_{N \times N}} \end{vmatrix} = D - CA^{-1}B \tag{16}$$

which shows that J_i are all $N \times N$ matrices and $\mathrm{G} = \mathrm{GL}(\mathrm{N}, \mathbb{C})$.

If we choose a seed solution J with $\det(J)$ is constant, by the Darboux transformation (13) we have $\det(\widetilde{J}) = \det(-\Lambda)\det(J)$ is also a constant. Applying the Jacobi's formula

$$\frac{d}{dt}\det A(t) = \mathrm{Tr}\left[\mathrm{adj}(A(t))\frac{dA(t)}{dt}\right] = \det A(t) \cdot \mathrm{Tr}(A(t)^{-1}\frac{dA(t)}{dt}) \tag{17}$$

to $\widetilde{J}$ and comparing with (11), we find that the gauge fields are all traceless. That is, any J-matrices (15) generated by a seed solution J_1 with constant determinant are belonging to $\mathrm{G} = \mathrm{SL}(\mathrm{N}, \mathbb{C})$ gauge theory. In fact, we will always set the seed solution J_1 to be identity matrix and only consider $N = 2, 3$ cases in the next sections.

3. ASDYM 1-Soliton and Multi-Soliton for G = SU(2)

Now we set the seed solution J_1 to be 2×2 identity matrix $(G = SL(2, \mathbb{C}))$ so that the linear system (12) reduces to

$$\begin{cases} L(\phi) = \partial_w \phi - (\partial_{\bar{z}} \phi)\zeta = 0 \\ M(\phi) = \partial_z \phi - (\partial_{\tilde{w}} \phi)\zeta = 0 \end{cases}. \tag{18}$$

For the split signature $(+, +, -, -)$, we can take the reality condition (7) and find a lovely solution of (18) as

$$\psi = \begin{pmatrix} ae^L & \bar{b}e^{-\bar{L}} \\ -be^{-L} & \bar{a}e^{\bar{L}} \end{pmatrix} \quad w.r.t. \text{ a specific spectral parameter } \Lambda = \begin{pmatrix} \lambda & 0 \\ 0 & \bar{\lambda} \end{pmatrix}, \tag{19}$$

where $L = \frac{1}{\sqrt{2}} \left[(\lambda\alpha + \beta)x^1 + (\lambda\beta - \alpha)x^2 + (\lambda\alpha - \beta)x^3 + (\lambda\beta + \alpha)x^4 \right]$, $a, b, \alpha, \beta, \lambda \in \mathbb{C}$.

After 1 iteration of the Darboux transformation, we obtain a candidate of 1-soliton solution :

$$J_2 = \begin{vmatrix} \psi & 1 \\ \psi\Lambda & \boxed{0} \end{vmatrix} = -\psi\Lambda\psi^{-1}$$

$$= \frac{-1}{\det(\psi)} \begin{pmatrix} \lambda|a|^2 e^{L+\bar{L}} + \bar{\lambda}|b|^2 e^{-(L+\bar{L})} & (\bar{\lambda} - \lambda)a\bar{b}e^{L-\bar{L}} \\ (\bar{\lambda} - \lambda)\bar{a}be^{-(L-\bar{L})} & \bar{\lambda}|a|^2 e^{L+\bar{L}} + \lambda|b|^2 e^{-(L+\bar{L})} \end{pmatrix}. \tag{20}$$

The resulting action density[4] of (20) is

$$\mathrm{Tr} F_{\mu\nu} F^{\mu\nu} = 8 \left[(\alpha\bar{\beta} - \bar{\alpha}\beta)(\lambda - \bar{\lambda}) \right]^2 \left(2\mathrm{sech}^2 X - 3\mathrm{sech}^4 X \right) \tag{21}$$
$$X = L + \bar{L} + \log(|a| / |b|)$$

We find that the distribution of action density is localized on a 3-dimensional hyperplane $X = 0$. Therefore, this kind of solitons can be interpreted as codimensional 1 soliton and we use the term soliton wall to distinguish them from the domain wall.

Now we can prepare n different solutions $\psi_i(\Lambda_i)$ of (18) as follows :

$$\psi_i = \begin{pmatrix} a_i e^{L_i} & \bar{b}_i e^{-\bar{L}_i} \\ -b_i e^{-L_i} & \bar{a}_i e^{\bar{L}_i} \end{pmatrix} \quad w.r.t. \text{ spectral parameters } \Lambda_i = \begin{pmatrix} \lambda_i & 0 \\ 0 & \bar{\lambda}_i \end{pmatrix},$$
$$L_i = \frac{1}{\sqrt{2}} \left[(\lambda_i\alpha_i + \beta_i)x^1 + (\lambda_i\beta_i - \alpha_i)x^2 + (\lambda_i\alpha_i - \beta_i)x^3 + (\lambda_i\beta_i + \alpha_i)x^4 \right], \tag{22}$$
$$a_i, b_i, \alpha_i, \beta_i, \lambda_i \in \mathbb{C}, \ i = 1, 2, \cdots, n.$$

After n-iterations of the Darboux transformation, we obtain a candidate of

n-soliton solution :

$$J_{n+1} = \begin{vmatrix} \psi_1 & \psi_2 & \cdots & \psi_n & 1 \\ \psi_1\Lambda_1 & \psi_2\Lambda_2 & \cdots & \psi_n\Lambda_n & 0 \\ \vdots & \vdots & \ddots & \vdots & \vdots \\ \psi_1\Lambda_1^n & \psi_2\Lambda_2^n & \cdots & \psi_n\Lambda_n^n & \boxed{0} \end{vmatrix}, \tag{23}$$

which satisfies the property[5,6]

$$J_{n+1}J_{n+1}^\dagger = J_{n+1}^\dagger J_{n+1} = \prod_{i=1}^{n} |\lambda_i|^2 I, \quad I : 2 \times 2 \text{ identity matrix.} \tag{24}$$

This fact implies that the gauge fields in 4-dimensional split signature

$$\begin{aligned} A_1^{(n+1)} = A_3^{(n+1)} = \tfrac{-1}{2}\left[(\partial_1 J_{n+1})J_{n+1}^{-1} + (\partial_3 J_{n+1})J_{n+1}^{-1} \right] \\ A_2^{(n+1)} = A_4^{(n+1)} = \tfrac{-1}{2}\left[(\partial_2 J_{n+1})J_{n+1}^{-1} + (\partial_4 J_{n+1})J_{n+1}^{-1} \right] \end{aligned} \tag{25}$$

are all anti-hermitian and therefore the gauge group is G = SU(2).

Now a natural question to ask is whether the n-soliton solution (23) gives rise to n intersecting soliton walls or not. Let us use a quite similar technique as mentioned in reference 7 to discuss the asymptotic action density of (23) rather than calculating the action density directly by (25). First of all, we fix an $I \in \{1, 2, ..., n\}$ and consider a comoving frame related to the I-th 1-soliton solution :

$$J_2^{(I)} = -\psi_n^{(I)}\Lambda_I(\psi_n^{(I)})^{-1}, \quad \psi_n^{(I)} = \begin{pmatrix} a_I e^{L_I} & \overline{b}_I\, e^{-\overline{L}_I} \\ -b_I e^{-L_I} & \overline{a}_I\, e^{\overline{L}_I} \end{pmatrix} \tag{26}$$

whose action density is

$$\mathrm{Tr}F_{\mu\nu}F^{\mu\nu\,(I)} = 8\left[(\alpha_I\overline{\beta}_I - \overline{\alpha}_I\beta_I)(\lambda_I - \overline{\lambda}_I) \right]^2 \left(2\mathrm{sech}^2 X_I - 3\mathrm{sech}^4 X_I \right) \\ X_I = L_I + \overline{L}_I + \log(|a_I|\,/\,|b_I|) \tag{27}$$

More precisely, we define $r := \sqrt{(x_1)^2 + (x_2)^2 + (x_3)^2 + (x_4)^2}$ and consider the asymptotic limit $r \to \infty$ such that

$$\begin{cases} X_I \text{ is a finite real number} \\ X_{i,i\neq I} \to \pm\infty \quad (i.e. \; \mathrm{Tr}F_{\mu\nu}F^{\mu\nu(i\neq I)} \to 0) \end{cases}. \tag{28}$$

By some mathematical techniques[5,6] of the quasideterminant, we find that

$$J_{n+1} \xrightarrow{r\to\infty} -\widetilde{\Psi}_n^{(I)}\Lambda_I(\widetilde{\Psi}_n^{(I)})^{-1}D_n^{(I)}, \; n \geq 2, \tag{29}$$

where $D_n^{(I)}$ is a constant matrix and doesn't affect the gauge fields, and

$$\widetilde{\Psi}_n^{(I)} := \begin{cases} (i) & \begin{pmatrix} \prod\limits_{i=1,i\neq I}^{n} (\lambda_I - \lambda_i)\, a_I e^{L_I} & \prod\limits_{i=1,i\neq I}^{n} (\overline{\lambda}_I - \lambda_i)\, \overline{b}_I\, e^{-\overline{L}_I} \\ -\prod\limits_{i=1,i\neq I}^{n} (\lambda_I - \overline{\lambda}_i)\, b_I e^{-L_I} & \prod\limits_{i=1,i\neq I}^{n} (\overline{\lambda}_I - \overline{\lambda}_i)\, \overline{a}_I\, e^{\overline{L}_I} \end{pmatrix} \\ & \text{as } X_{i,i\neq I} \to +\infty \\[2mm] (ii) & \begin{pmatrix} \prod\limits_{i=1,i\neq I}^{n} (\lambda_I - \overline{\lambda}_i)\, a_I e^{L_I} & \prod\limits_{i=1,i\neq I}^{n} (\overline{\lambda}_I - \overline{\lambda}_i)\, \overline{b}_I\, e^{-\overline{L}_I} \\ -\prod\limits_{i=1,i\neq I}^{n} (\lambda_I - \lambda_i)\, b_I e^{-L_I} & \prod\limits_{i=1,i\neq I}^{n} (\overline{\lambda}_I - \lambda_i)\, \overline{a}_I\, e^{\overline{L}_I} \end{pmatrix} \\ & \text{as } X_{i,i\neq I} \to -\infty \end{cases} \qquad (30)$$

Comparing the asymptotic n-solion solution (29), (30) with I-th 1-soliton solution (26), we find that $\widetilde{\Psi}_n^{(I)}$ is in the same form as $\psi_n^{(I)}$ up to constant factors. These factors lead to a position shift from the principal peak of action density (27), called phase shift. More specifically, the action density of n-soliton solution in the asymptotic region

$$\mathrm{Tr} F_{\mu\nu} F^{\mu\nu} \overset{r\to\infty}{\longrightarrow} 8\left[(\alpha_I \overline{\beta}_I - \overline{\alpha}_I \beta_I)(\lambda_I - \overline{\lambda}_I)\right]^2 \left(2\mathrm{sech}^2 \widetilde{X}_I - 3\mathrm{sech}^4 \widetilde{X}_I\right) \quad (31)$$

behaves like the action density of I-th 1-soliton (27), where $\widetilde{X}_I := X_I + \Delta_I$ and the phase shift

$$\Delta_I = \sum_{i=1,i\neq I}^{n} \varepsilon_i^{(\pm)} \log\left|\frac{\lambda_I - \lambda_i}{\lambda_I - \overline{\lambda}_i}\right|, \qquad \begin{cases} \varepsilon_i^{(+)} := +1, & X_{i,i\neq I} \to +\infty \\ \varepsilon_i^{(-)} := -1, & X_{i,i\neq I} \to -\infty \end{cases} \quad (32)$$

is real-valued. Since I is any positive integer from 1 to n, and for every I the n-soliton solution gives rise to a soliton wall in the asymptotic region, we can conclude that the n-soliton solution can be interpreted as n intersecting soliton walls in the entire region. Furthermore, the n intersecting soliton walls can be embedded into G = SU(2) gauge theory in 4-dimensional split signature and therefore they could be interpreted as n intersecting branes in open N = 2 string theories.

4. Reduction to (1+1)-dimensional real space

To make the discussion more clearly, let us take 2-dimensional spacetime and 3-soliton scattering for instance. We can impose the condition $x^1 = t$, $x^2 = x^4 = 0$, $x^3 = x$ on the spacetime coordinates such that

$$L_i = \frac{1}{\sqrt{2}}\left((\lambda_i \alpha_i + \beta_i)t + (\lambda_i \alpha_i - \beta_i)x\right), \; i = 1, 2, 3 \quad (33)$$

and $X_i = L_i + \overline{L}_i + \log|a_i/b_i|$. Then we fix an $I \in \{1,2,3\}$ and choose a complex number ℓ_I such that $L_I = \ell_I$ which implies

$$L_i = \left(\frac{\lambda_i\alpha_i - \beta_i}{\lambda_I\alpha_I - \beta_I}\right)\ell_I + \sqrt{2}\left(\frac{\lambda_i\alpha_i\beta_I - \lambda_I\alpha_I\beta_i}{\lambda_I\alpha_I - \beta_I}\right)t, \quad i \neq I. \quad (34)$$

Now the setup of the comoving frame related to the I-th 1-soliton is completed because

$$\begin{cases} X_I = \text{finite} \\ X_{i,i\neq I} \to \pm\infty \ \text{ or } \ \mp\infty \end{cases} \text{when } t \to \pm\infty \quad (35)$$

This setup implies that the action density of i-th 1-soliton ($i \neq I$) behaves as $\text{Tr}F_{\mu\nu}F^{\mu\nu\,(i\neq I)} \sim 2\text{sech}^2 X_i - 3\text{sech}^4 X_i \to 0$ as $t \to \pm\infty$. On the other hand, the action density of 3-soliton in the asymptotic region behaves as

$$\text{Tr}F_{\mu\nu}F^{\mu\nu} \sim 2\text{sech}^2\widetilde{X}_I - 3\text{sech}^4\widetilde{X}_I, \ \widetilde{X}_I = \begin{cases} X_I + \Delta_I^{(+)} & \text{as } t \to +\infty \\ X_I + \Delta_I^{(-)} & \text{as } t \to -\infty \end{cases}. \quad (36)$$

In fact, $\Delta_I^{(-)} = -\Delta_I^{(+)}$ and it depends on $2^{3-1} = 4$ choices of asymptotic regions. For example one of the choices is

$$\Delta_1^{(+)} = -\Delta_1^{(-)} = -\log\left|\frac{\lambda_1 - \lambda_2}{\lambda_1 - \overline{\lambda}_2}\right| - \log\left|\frac{\lambda_1 - \lambda_3}{\lambda_1 - \overline{\lambda}_3}\right|, \quad (37)$$

$$\Delta_2^{(+)} = -\Delta_2^{(-)} = +\log\left|\frac{\lambda_2 - \lambda_1}{\lambda_2 - \overline{\lambda}_1}\right| - \log\left|\frac{\lambda_2 - \lambda_3}{\lambda_3 - \overline{\lambda}_3}\right|, \quad (38)$$

$$\Delta_3^{(+)} = -\Delta_3^{(-)} = +\log\left|\frac{\lambda_3 - \lambda_1}{\lambda_3 - \overline{\lambda}_1}\right| + \log\left|\frac{\lambda_3 - \lambda_2}{\lambda_3 - \overline{\lambda}_2}\right|. \quad (39)$$

5. An example of ASDYM 1-Soliton for G = SU(3) and Multi-Soliton scattering

If we consider the linear system (18) for $G = SL(3,\mathbb{C})$, these is a special class of solutions $\psi_i(\Lambda_i)$ that give rise to soliton walls[6] and behave like the $G = SU(2)$ cases as well. Firstly, the candidate of $G = SL(3,\mathbb{C})$ 1-soliton solution can be constructed by 1-iteration of the Darboux transformation

$$J_2^{(i)} = -\psi_i\Lambda_i\psi_i^{-1}, \quad \begin{cases} \psi_i := \begin{pmatrix} a_ie^{L_i} & \overline{b}_ie^{-\overline{L}_i} & 0 \\ -b_ie^{-L_i} & \overline{a}_ie^{\overline{L}_i} & c_ie^{-L_i} \\ 0 & -\overline{c}_ie^{-\overline{L}_i} & a_ie^{L_i} \end{pmatrix}, \\ \Lambda_i := \begin{pmatrix} \lambda_i & 0 & 0 \\ 0 & \overline{\lambda}_i & 0 \\ 0 & 0 & \lambda_i \end{pmatrix} \end{cases} \quad (40)$$

$$L_i = \frac{1}{\sqrt{2}}\left[(\lambda_i\alpha_i + \beta_i)x^1 + (\lambda_i\beta_i - \alpha_i)x^2 + (\lambda_i\alpha_i - \beta_i)x^3 + (\lambda_i\beta_i + \alpha_i)x^4\right],$$
$$a_i, b_i, c_i, \alpha_i, \beta_i, \lambda_i \in \mathbb{C}, \ i = 1,2,...,n.$$

By direct calculation, we obtain the action density of i-th 1-soliton as

$$\text{Tr}F_{\mu\nu}F^{\mu\nu\,(i)} = 8\left[(\alpha_i\overline{\beta}_i - \overline{\alpha}_i\beta_i)(\lambda_i - \overline{\lambda}_i)\right]^2\left(2\text{sech}^2 X_i - 3\text{sech}^4 X_i\right)$$
$$X_i = L_i + \overline{L}_i + \tfrac{1}{2}\log\left[|a_i|^2/(|b_i|^2 + |c_i|^2)\right] \quad, \quad (41)$$

which can be interpreted as i-th soliton wall as discussed in $G = SU(2)$ case. Substituting ψ_i of (40) into (23), we get a version of n-soliton J_{n+1} for $G = SL(3,\mathbb{C})$. By some mathematical techniques[6] of the quasideterminant and considering a comoving frame related to the I-th 1-soliton, we find that

$$J_{n+1} \xrightarrow{r\to\infty} \widetilde{J}_{n+1}^{(I)} =: -\widetilde{\Psi}_n^{(I)}\Lambda_I(\widetilde{\Psi}_n^{(I)})^{-1}D_n^{(I)}, \quad n \geq 2, \qquad (42)$$

where $D_n^{(I)}$ is a constant matrix and $\widetilde{\Psi}_n^{(I)}$ is in the form of

$$\widetilde{\Psi}_n^{(I)} = \begin{pmatrix} A_I e^{L_I} & \overline{B}_I e^{-\overline{L}_I} & 0 \\ -B_I e^{-L_I} & \overline{A}_I e^{\overline{L}_I} & C_I e^{-L_I} \\ 0 & -\overline{C}_I e^{-\overline{L}_I} & A_I e^{L_I} \end{pmatrix}, \quad \begin{cases} A_I = \prod\limits_{i=1,i\neq I}^{n}(\lambda_I - \lambda_i^{(\pm)})\,a_I \\ B_I = \prod\limits_{i=1,i\neq I}^{n}(\lambda_I - \lambda_i^{(\mp)})\,b_I \\ C_I = \prod\limits_{i=1,i\neq I}^{n}(\lambda_I - \lambda_i^{(\mp)})\,c_I \\ (\lambda_i^{(+)}, \lambda_i^{(-)}) := (\lambda_i, \overline{\lambda}_i) \end{cases} \quad (43)$$

Comparing (43) with (40) and (41), we can conclude that the action density of n-soliton in the asymptotic region

$$\text{Tr}F_{\mu\nu}F^{\mu\nu} \xrightarrow{r\to\infty} 8\left[(\alpha_I\overline{\beta}_I - \overline{\alpha}_I\beta_I)(\lambda_I - \overline{\lambda}_I)\right]^2\left(2\text{sech}^2\widetilde{X}_I - 3\text{sech}^4\widetilde{X}_I\right) \quad (44)$$

behaves like the action density of I-th 1-soliton, where $\widetilde{X}_I := X_I + \Delta_I$ and the phase shift is

$$\Delta_I = \frac{1}{2}\log\left[\frac{\prod\limits_{i=1,i\neq I}^{n}\left|\lambda_I - \lambda_i^{(\pm)}\right|^2(|b_I|^2 + |c_I|^2)}{\prod\limits_{i=1,i\neq I}^{n}\left|(\lambda_I - \lambda_i^{(\mp)})\right|^2|b_I|^2 + \prod\limits_{i=1,i\neq I}^{n}\left|(\lambda_I - \lambda_i^{(\mp)})\right|^2|c_I|^2}\right] \quad (45)$$

which is real-valued and depends on 2^{n-1} choices of $(\lambda_i^{(+)}, \lambda_i^{(-)}) := (\lambda_i, \overline{\lambda}_i)$. Therefore, the n-soliton can be interpreted as n-intersecting soliton walls for $G = SL(3,\mathbb{C})$. On the other hand, the asymptotic form $\widetilde{J}_{n+1}^{(I)}$ of the n-soliton solution J_{n+1} satisfies the property[6]

$$\widetilde{J}_{n+1}^{(I)\,\dagger}\widetilde{J}_{n+1}^{(I)} = \widetilde{J}_{n+1}^{(I)}\widetilde{J}_{n+1}^{(I)\,\dagger} = \prod\limits_{i=1}^{n}|\lambda_i|^2 I_{3\times 3} \qquad (46)$$

which implies that the gauge fields given by $\widetilde{J}_{n+1}^{(I)}$ are all anti-hermitian. Therefore for each single soliton wall in the asymptotic region, the gauge group is in fact $G = SU(3)$.

6. Conclusion

In this paper, we found a class of ASDYM 1-solitons in 4-dimensional split signature for G = SU(2) and SU(3), respectively. The resulting action densities are in the same form as $\mathrm{Tr} F_{\mu\nu} F^{\mu\nu} \propto \left(2\mathrm{sech}^2 X - 3\mathrm{sech}^4 X\right)$ which can be interpreted as the soliton walls. After n-iterations of the Darboux transformation, we obtain the ASDYM n-soliton and interpret it as n intersecting soliton walls for G = SU(2) and SL(3, $\mathbb{C}$), respectively. This fact is a well-known feature for the KdV multi-solitons, but a new insight for the ASDYM multi-solitons. On the other hand, the n intersecting soliton walls can be embedded into G = SU(2) gauge theory and hence they could be interpreted as n intersecting branes in N = 2 open string theories. Therefore, to understand the role of such physical objects that play in N = 2 open string theories would be an interesting future work, and the relationship between N = 2 open string theories and lower-dimensional integrable systems is also worth studying.

References

1. Y. Brihaye, D. B. Fairlie, J. Nuyts and R. G. Yates, J. Math. Phys. **19**, pp.2528-2532 (1978).
2. I. Gelfand and V. Retakh, Funct. Anal. Appl. **25**, pp.91-102 (1991).
3. C. R. Gilson, M. Hamanaka, S. C. Huang and J. J. C. Nimmo, J. Phys. A: Math. Theor. **53**, 404002 (17pp) (2020).
4. M. Hamanaka and S.C. Huang, JHEP **10**, 101 (2020).
5. M. Hamanaka and S.C. Huang, JHEP **01**, 039 (2022).
6. S.C. Huang, "On soliton solutions of the anti-self-dual Yang-Mills equations from the perspective of integrable systems", [arXiv:2112.10702].
7. V. B. Matveev and M. A. Salle, *Darboux Transformations and Solitons*, (Springer-Verlag, 1991).
8. L. J. Mason and N. M. Woodhouse, *Integrability, Self-Duality, and Twistor Theory* (Oxford UP, 1996).
9. R. M. Miura, C. S. Gardner, and M. D. Kruskal, J. Math. Phys. **9**, pp.12041209 (1968).
10. J. J. C. Nimmo, C. R. Gilson and Y. Ohta, Theor. Math. Phys. **122**, pp.239-246 (2000).
11. H. Ooguri and C. Vafa, Nucl. Phys. B **361** (2), pp.469-518 (1991); Nucl. Phys. B **367** (1), pp.83-104 (1991).
12. R. S. Ward, Phil. Trans. Roy. Soc. Lond. A **315**, pp.451-457 (1985).
13. C. N. Yang, Phys. Rev. Lett. **38**, pp.1377-1379 (1977).

Wall-crossing of TBA equations and WKB periods for the higher order ODE

Katsushi Ito[a], Takayasu Kondo[b]

Department of Physics, Tokyo Institute of Technology, Tokyo, 152-8551, Japan
[a] *E-mail: ito@th.phys.titech.ac.jp*
[b] *t.kondo@th.phys.titech.ac.jp*

Hongfei Shu

Beijing Institute of Mathematical Sciences and Applications (BIMSA), Beijing, 101408, China
Yau Mathematical Sciences Center (YMSC), Tsinghua University, Beijing, 100084, China
E-mail: shuphy124@gmail.com

We first study the WKB analysis of the third order ODE, which can be regarded as the quantized Seiberg-Witten curve of the (A_2, A_N)-type Argyres-Douglas theory in the Nekrasov-Shatashvili limit of Omega background. We then derive thermodynamic Bethe ansatz (TBA) equations satisfied by the Y-functions from the solutions of the ODE, and identify the Y-function with the WKB period. For the (A_2, A_2)-type ODE, we study the process of wall-crossing of the TBA equation from the minimal chamber to the maximal chamber.

Keywords: TBA equations, WKB period, Wall-crossing, ODE/IM correspondence.

1. Introduction

The WKB period of the Schrödinger equation is expanded as the formal asymptotic series in the Plank constant, which thus needs to be Borel resummed. Recently, a connection between the exact WKB period and the thermodynamic Bethe ansatz (TBA) equations of the quantum integral model has been noticed for the Schrödinger equation with arbitrary polynomial potential[1,2]. In particular, the Y-function of the TBA equations is identified to the exponential of the Borel resummed WKB period, which share the same asymptotic behaviors and the discontinuity on the complex $\hbar$-plane. The TBA equations thus provide a solution to the Voros' Riemann-Hilbert problem of the exact WKB periods[3]. Moreover, the TBA equations together with the exact quantization condition lead to the exact energy spectrum of the one-dimensional Quantum Mechanics[1,4]. The Schrödinger-type equation with polynomial potential also appears as the quantized Seiberg-Witten curve of the Argyres-Douglas (AD) theories in the Nekrasov-Shatashvili (NS) limit of the Ω background[5-7], where the non-zero parameter of the Ω-background plays the role of Planck constant. The TBA equations satisfied by the WKB period (quantum period) thus becomes very useful to compute the exact BPS spectrum, whose wall-crossing was observed when the moduli space parameters are varied. Due to the wall-crossing, the form of the TBA equations also changed, which we will denote by the wall-crossing of the TBA equations.

To study the exact BPS spectrum of the higher rank AD theories, it is important to explore the relations between the exact WKB periods of the higher order ODE and the TBA equations in the quantum integrable model, whose wall-crossing will be useful to understand the wall-crossing phenomena of more general non-perturbative supersymmetric gauge theories. In[8], we have focused on the higher order ODE with quadratic potential, which can be regarded as the quantized SW curve of the (A_r, A_1)-type AD theory. The corresponding TBA equations are found to be (A_r, A_1) type TBA equations. Moreover, the wall-crossing of TBA equations was observed for the third order ODE with cubic potential. The wall-crossing of the TBA equations for more general third order ODE has been studied in[9]. For the monomial type potential, the TBA equations obtained from the (A_2, A_2)-type and (A_2, A_3)-type ODE become the D_4-type and E_6-type TBA equations, respectively.

This proceeding is organized as follows. In section 2, we study the WKB analysis of the third order ODE and introduce the Borel resummation of the WKB period. In section 3, we derived the TBA equations satisfied by the Y-functions, and identify them with the exact WKB periods in the minimal chamber. In section 4, we study the wall-crossing of the TBA equations. The TBA equation in the maximal chamber will be constructed. The section 5 is is devoted to conclusions and discussion.

2. WKB analysis of the third order ODE

We consider the third order ODE on the complex plane:

$$\left(\epsilon^3 \frac{d^3}{dx^3} + p(x) \right) \psi(x) = 0, \quad p(x) = u_0 x^{N+1} + u_1 x^N + \cdots + u_{N+1} \quad (1)$$

where ϵ and u_i $(i = 0, \cdots, N+1)$ are complex parameters. This ODE can be regarded as the quantized SW curve of the (A_2, A_N)-type AD theory in the NS limit of the Ω-background. We will denote this ODE by (A_2, A_N)-type ODE in this paper.

The WKB analysis begins by imposing the following ansatz of the wave function

$$\psi(x) = \exp\left(\frac{1}{\epsilon} \int^x P(x')dx'\right), \qquad (2)$$

where $P(x)$ satisfies the Ricatti equation

$$p(x) + P^3 + 3\epsilon PP' + \epsilon^2 P^{(2)} = 0. \qquad (3)$$

The solution can be expanded in power series of ϵ

$$P(x) = \sum_{n=0}^{\infty} \epsilon^n p_n(x), \qquad (4)$$

which can be determined recursively by

$$p_0 = (-p)^{\frac{1}{3}},$$

$$p_n = -\frac{1}{3p_0^2}\left[p_0 \sum_{i=1}^{n-1} p_{n-i}p_i + \sum_{i=1}^{n-1}\sum_{j=0}^{n-i} p_{n-i-j}p_ip_j \right.$$
$$\left. + 3\sum_{i=0}^{n-1} p_{n-1-i}p_i' + p_{n-2}''\right], \quad n \geq 1. \qquad (5)$$

It is natural to regard Pdx as the meromorphic differential on the WKB curve Σ

$$y^3 + p(x) = 0, \qquad (6)$$

which is nothing but the SW curve of (A_2, A_N) AD theory. We introduce the period of Pdx along the one-cycle γ:

$$\Pi_\gamma(\epsilon) = \int_\gamma P(x)dx = \sum_{n=0}^{\infty} \epsilon^n \Pi_\gamma^{(n)}, \qquad (7)$$

which will be denoted by the WKB period. Since the basis of the meromorphic differentials on the WKB curve can be generated by

$$\partial_{u_i} y^a dx = -\frac{a}{3} \frac{x^{N+1-i}}{y^{3-a}} dx, \quad a = 1, 2, \qquad (8)$$

we can obtain the quantum correction of the WKB periods by acting the differential operators of u_i on the periods of y^a:

$$\Pi_\gamma^{(n)} = \sum_{a=1}^{2} \mathcal{O}_a^{(n)} \hat{\Pi}_{a\gamma}, \qquad (9)$$

where

$$(\hat{\Pi}_a)_\gamma = \int_\gamma y^a dx. \tag{10}$$

The operator is called Picard-Fuchs operators, whose details can be found in [8,9].

Since the WKB periods is an asymptotic formal series in ϵ, we need to perform the Borel transform

$$\mathcal{B}[\Pi_\gamma](\xi) = \sum_{n \geq 0} \frac{1}{n!} \Pi_\gamma^{(n)} \xi^n, \tag{11}$$

which Laplace transform along the direction φ is called the Borel resummation

$$s_\varphi(\Pi_\gamma)(\epsilon) = \frac{1}{\epsilon} \int_0^{\infty e^{i\varphi}} e^{-\xi/\epsilon} \mathcal{B}[\Pi_\gamma](\xi) d\xi. \tag{12}$$

The WKB period is Borel summable when the Borel transform converges. There arises a discontinuity for the resummed WKB period

$$\begin{aligned} \mathrm{disc}_\varphi \Pi_\gamma(\epsilon) &= s_{\varphi^+}(\Pi_\gamma)(\epsilon) - s_{\varphi^-}(\Pi_\gamma)(\epsilon)) \\ &= \lim_{\delta \to 0_+} \left(s(\Pi_\gamma)(e^{i\varphi + i\delta}\epsilon) - s(\Pi_\gamma)(e^{i\varphi - i\delta}\epsilon) \right), \end{aligned} \tag{13}$$

when $\mathcal{B}[\Pi_\gamma]$ is singular along the direction φ. By using the Borel-Pade technique, we are able to compute the singularity structure of the Borel transform numerically. In Fig.1, we plot the singularity structure of the Borel transform of the WKB periods for the potential $p(x) = -x^3 + 7x + 6$, which will be compared to the TBA results later.

3. TBA/WKB correspondence of the third order ODE

In this section, we derive the TBA equation derived from the solutions of the ODE, and identify the Y-function with the WKB periods.

The ODE (1) is invariant under the rotation:

$$x \to \omega^{-1} x, \quad u_i \to \omega^{-i} u_i, \qquad i = 0, 1, \dots, N + 1 \tag{14}$$

with $\omega = e^{2\pi i/(N+4)}$. Since the solution to the ODE is changed, the rotation is very useful to generate the solutions of the ODE

$$\psi(\omega^{-1} x, \{\omega^{-i} u_i\}; \epsilon) = \psi(x, \{u_i\}; e^{2\pi i/3} \epsilon), \tag{15}$$

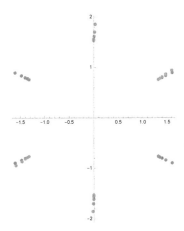

Fig. 1. The singularity structure of the Borel transformed WKB periods $\Pi_{\gamma_{1,1}}$ (blue) and $\Pi_{\gamma_{1,1}}$ (yellow). The Borel-Padé technique is applied to order ϵ^{160} terms of the formal power series computed by using the Picard-Fuchs operators. Here the potential is $p(x) = -x^3 + 7x + 6$. The figure was first presented in[9].

where we have rewrite the rotation in terms of ϵ. At the infinity of the real and positive axis, the subdominant solution of the ODE behaves as

$$\phi_0(x, \{u_i\}; \epsilon)$$

$$\sim \frac{\epsilon}{i\sqrt{3}} x^{-\frac{N+1}{3}} \exp(-\frac{1}{\epsilon}\frac{3}{N+4}x^{\frac{N+4}{3}}), \quad |x| \to \infty, \quad |\arg x| < \frac{\pi}{N+4}, \quad (16)$$

where we have set $u_0 = 1$ without loss of generality. The rotated solution ϕ_k defined by

$$\phi_k(x, \{u_i\}; \epsilon) = \phi_0(x, \{u_i\}; e^{\frac{2\pi i}{3}k}\epsilon) \quad (17)$$

are the subdominant solution in the sector

$$\mathcal{S}_k = \left\{ x \in \mathbb{C}; |\arg x| - \frac{2\pi k}{N+4} < \frac{\pi}{N+4} \right\}. \quad (18)$$

Using these solutions, we introduce the T-functions

$$T_{0,k} = W[\phi_{-1}, \phi_0, \phi_1]^{[-k-1]} = 1, \quad T_{1,k} = W[\phi_{-1}, \phi_0, \phi_{k+1}]^{[-k]},$$
$$T_{2,k} = W[\phi_0, \phi_{k+1}, \phi_{k+2}]^{[-k-1]}, \quad T_{3,k} = W[\phi_k, \phi_{k+1}, \phi_{k+2}]^{[-k]} = 1, \quad (19)$$

where $f^{[k]}(\{u_i\}; \epsilon) = f(\{u_i\}; e^{\frac{\pi i}{3}k}\epsilon)$ and $W[\cdot]$ is the Wronskian

$$W[f_{k_1}, f_{k_2}, f_{k_3}] = \det \begin{pmatrix} f_{k_1} & f_{k_2} & f_{k_3} \\ \partial_x f_{k_1} & \partial_x f_{k_2} & \partial_x f_{k_3} \\ \partial_x^2 f_{k_1} & \partial_x^2 f_{k_2} & \partial_x^2 f_{k_3} \end{pmatrix}. \quad (20)$$

48

We introduce the Y-function by the cross ratios of the T-functions

$$Y_{a,k} = \frac{T_{a-1,k}T_{a+1,k}}{T_{a,k-1}T_{a,k+1}}, \qquad a = 1,2, \quad k = 1,\ldots,N. \tag{21}$$

By using the identities of the Wronskian, it is to find these Y-function satisfy the (A_2, A_N)-type Y-system[10]:

$$Y_{a,k}^{[+1]}Y_{a,k}^{[-1]} = \frac{(1 + Y_{a-1,k})(1 + Y_{a+1,k})}{(1 + Y_{a,k-1}^{-1})(1 + Y_{a,k+1}^{-1})} \tag{22}$$

Using the WKB approximation of the solutions ϕ_k and the Stokes graphs, we find the asymptotics of the Y-function are determined by the WKB period. Together with the discontinuity structure, we find

$$\log Y_{a,k} = \left[\tfrac{1}{\epsilon}\Pi_{\hat\gamma_{a,k}}\right]^{[a-k]}, \quad \hat\gamma_{a,k} = \gamma_{2-k,k} + \cdots + \gamma_{a+1-k,k}, \\ a = 1,2, \quad k = 1,\ldots,N, \tag{23}$$

where $\gamma_{l,k}$ is the one-cycle encircling the branch points x_{k-1} anticlockwise and x_k clockwise, respectively, on l-th and $(l+1)$-th sheets of the WKB curve. In the following of this section, we will test this relation numerically.

Based on the asymptotics of the Y-functions, it is easy to convert the Y-system into TBA equations

$$\log Y_{1,k}(\theta - i\phi_k) = |m_{1,k}|e^\theta + K \star \overline{L}_{1,k} - K_{k,k-1} \star \overline{L}_{1,k-1} - K_{k,k+1} \star \overline{L}_{1,k+1}, \tag{24}$$

where $\star$ denotes the convolution, $m_{a,k} = e^{\frac{\pi i}{3}(k-a)}\Pi_{\hat\gamma_{a,k}}^{(0)}$ and $\phi_i = \arg(m_{1,k})$. The kernel is defined by

$$K(\theta) = \frac{1}{2\pi}\frac{4\sqrt{3}\cosh\theta}{1 + 2\cosh 2\theta}, \quad K_{k_1,k_2}(\theta) = K(\theta - i(\phi_{k_1} - \phi_{k_2})). \tag{25}$$

At small $\epsilon = e^{-\theta}$, we can expand the TBA equations by

$$\log Y_{1,k}(\theta) = m_{1,k}e^\theta + \sum_{n=1}^\infty m_{1,k}^{(n)}e^{-n\theta}, \tag{26}$$

where the coefficient $m_{1,k}^{(n)}$ can be computed by using the solutions of the TBA equations. In table 3, we compare the ϵ-expansion of the two hand sides of (23) for the potential $p(x) = -x^3 + 7x + 6$.

The ϵ-expansion of the two hand sides of (23) for $p(x) = -x^3 + 7x + 6$[9].

n	$\Pi^{(n)}_{\hat{\gamma}_{1,1}}$	$m^{(n-1)}_{1,1}$
2	$0.2172157436i$	$0.2172157436i$
6	$-1.519567945i$	$-1.519567945i$
8	$-20.48661777i$	$-20.48661776i$
12	$20065.20970i$	$20065.20605i$
14	$1160395.676i$	$1160393.422i$

Moreover, we can express the left hand side of (23) by using the TBA equations:

$$\log Y_{1,1}(\theta) = m_{1,1}e^{\theta} + \int_{-\infty}^{\infty} d\theta' K(\theta - \theta' + i\phi_1)\overline{L}_{1,1}(\theta')$$

$$- \int_{-\infty}^{\infty} d\theta' K(\theta - \theta' + i\phi_2)\overline{L}_{1,2}(\theta') + \cdots,$$

$$\log Y_{1,2}(\theta - \frac{\pi i}{3}) = m_{1,2}e^{\theta - \frac{\pi i}{3}} + \int_{-\infty}^{\infty} d\theta' K(\theta - \theta' + i\phi_2 - \frac{\pi i}{3})\overline{L}_{1,2}(\theta')$$

$$- \int_{-\infty}^{\infty} d\theta' K(\theta - \theta' + i\phi_1 - \frac{\pi i}{3})\overline{L}_{1,1}(\theta') + \cdots,$$

$$(27)$$

When $\arg(m_{1,1}) = \arg(m_{1,2}) = \frac{\pi}{2}$, i.e. the case studied in Fig.1 and table 3, we find the poles of the kernel of the TBA equations are located along the directions $\theta = -\frac{5\pi}{6}, -\frac{\pi}{6}, \frac{\pi}{6}, \frac{5\pi}{6}$ for $\log Y_{1,1}(\theta)$ and $\theta = -\frac{\pi}{2}, -\frac{5\pi}{6}, \frac{\pi}{6}, \frac{\pi}{2}$ for $\log Y_{1,2}(\theta - \frac{\pi i}{3})$. Comparing with the singularity structure of Fig.1, we find these poles are nothing but the discontinuity of the Borel resummed WKB period of the right hand side of (23).

4. Wall-crossing of the TBA equations

In this section, we study the wall-crossing of the TBA equations for the case of (A_2, A_2) ODE. We parameterize the zeros, $x_0(t)$, $x_1(t)$ and $x_2(t)$, of the potential in the following way

$$x_0(t) = 3 - t, \quad x_1(t) = -1 + \sqrt{3}it, \quad x_2(t) = -2 + t - \sqrt{3}it, \quad 0 \le t \le 1$$
$$(28)$$

and denote the potential by $p(x,t) = (x - x_0(t))(x - x_1(t))(x - x_2(t))$. For a given value t, we can compute the classical periods and mass of the the corresponding Y-functions. We find that the pole of the kernel in TBA equations (24) are located at

$$t = 0.162117..., \qquad \phi_2 - \phi_1 = \frac{\pi}{3}, \qquad \Im\left(\frac{\Pi_{\gamma_{3,2}}^{(0)}}{\Pi_{\gamma_{1,1}}^{(0)}}\right) = 0, \qquad (29)$$

$$t = 0.397459..., \qquad \phi_2 - \phi_1 = \frac{2\pi}{3}, \qquad \Im\left(\frac{\Pi_{\gamma_{2,2}}^{(0)}}{\Pi_{\gamma_{1,1}}^{(0)}}\right) = 0, \qquad (30)$$

which are nothing but the marginalstability walls of the (A_2, A_2) AD theory. Crossing the first wall at $\phi_2 - \phi_1 = \frac{\pi}{3}$, one needs to pick up the contribution of the pole and modify the TBA equations:

$$\log Y_{1,1}(\theta - i\phi_1) = |m_{1,1}|e^\theta + K \star \overline{L}_{1,1} - K_{1,2} \star \overline{L}_{1,2} - L_{1,2}(\theta - \frac{\pi i}{3} - i\phi_1),$$

$$\log Y_{1,2}(\theta - i\phi_2) = |m_{1,2}|e^\theta - K_{2,1} \star \overline{L}_{1,1} + K \star \overline{L}_{1,2} - L_{1,1}(\theta + \frac{\pi i}{3} - i\phi_2)$$

$$(31)$$

It is more convenient to introduce the new Y-function

$$Y_{1,1}^{(1)}(\theta) = Y_{1,1}(\theta)\left(1 + \frac{1}{Y_{1,2}(\theta - \frac{\pi i}{3})}\right), \quad Y_{1,2}^{(1)}(\theta) = Y_{1,2}(\theta)\left(1 + \frac{1}{Y_{1,1}(\theta + \frac{\pi i}{3})}\right),$$

$$Y_{12}^{(1)}(\theta) = \frac{1 + \frac{1}{Y_{1,2}(\theta - \frac{\pi i}{3})} + \frac{1}{Y_{1,1}(\theta)}}{\frac{1}{Y_{1,1}(\theta)Y_{1,2}(\theta - \frac{\pi i}{3})}},$$

$$(32)$$

such that the modified TBA equations can be rewritten as a three-TBA equations system

$$\log Y_{1,1}^{(1)}(\theta - i\phi_1) = |m_{1,1}|e^\theta + K \star \overline{L}_{1,1}^{(1)} - K_{1,2} \star \overline{L}_{1,2}^{(1)} + K_{1,12}^- \star \overline{L}_{12}^{(1)},$$

$$\log Y_{1,2}^{(1)}(\theta - i\phi_2) = |m_{1,2}|e^\theta + K \star \overline{L}_{1,2}^{(1)} - K_{2,1} \star \overline{L}_{1,1}^{(1)} - K_{2,12}^- \star \overline{L}_{12}^{(1)}, \quad (33)$$

$$\log Y_{12}^{(1)}(\theta - i\phi_{12}) = |m_{12}|e^\theta + K \star \overline{L}_{12}^{(1)} + K_{12,1}^+ \star \overline{L}_{1,1}^{(1)} - K_{12,2}^+ \star \overline{L}_{1,2}^{(1)}.$$

These new Y-functions are identified to the WKB periods in the following way

$$\log Y_{1,1}^{(1)}(\theta) = e^\theta \Pi_{\hat{\gamma}_{1,1}}, \quad \log Y_{1,2}^{(1)}(\theta) = e^{\frac{\pi i}{3}} e^\theta \Pi_{\hat{\gamma}_{1,2}}(\theta + \frac{\pi i}{3}),$$

$$\log Y_{12}^{(1)}(\theta) = e^\theta \Pi_{\hat{\gamma}_{1,1} + \hat{\gamma}_{1,2}}(\theta),$$

$$(34)$$

which have been tested numerically. Moreover, we can cross the second wall at $\phi_2 - \phi_1 = 2\pi/3$ and arrive at the maximal chamber, where four-TBA equations system of $Y_{1,1}^{(2)}, Y_{1,2}^{(2)}, Y_{12}^{(2)}$ and $Y_{12}^{(2)}$ were found. At the monomial potential, i.e. $p(x, t = 1) = x^3 - 8$, we can compute the classical periods and the masses of the Y-function, which leads to a symmetry between the Y-functions

$$\log Y_{1,1}^{(2)}(\theta - i\phi_1) = \log Y_{1,2}^{(2)}(\theta - i\phi_1 - \pi i) = \log Y_{12}^{(2)}(\theta - i\phi_1 - \frac{\pi i}{3}). \quad (35)$$

The TBA equations at the monomial potential thus become a two-TBA equations system, which has the same form as the (symmetric-) D_4-type TBA equations [10,11].

5. Conclusions and discussion

In this paper, we have found the correspondence between the WKB periods of the third order ODE and the Y-functions satisfying the TBA equation in the quantum integrable model. We have studied the wall-crossing of the TBA equations on the moduli space. For the (A_2, A_2)-type ODE, we found the TBA equations at the monomial point become the D_4-type TBA equations. Our observation can be regarded as the duality between (A_2, A_2)-type and D_4-type AD theory at the quantum level. The TBA equations of different integrable models are connected through the wall-crossing. It would be interesting to see how these integrable models are unified at the IM side.

Acknowledgements

We would like to thank Davide Fioravanti, Daniele Gregori, Kohei Kuroda, Yongchao Lü, Hao Ouyang, Marco Rossi, and Dan Xie for useful discussions. The work of K.I. is supported in part by Grant-in-Aid for Scientific Research 21K03570, 18K03643, and 17H06463 from Japan Society for the Promotion of Science (JSPS). The work of H.S. is supported by the grant "Exact Results in Gauge and String Theories" from the Knut and Alice Wallenberg foundation. H.S. would like to thank Jilin University for their (online) hospitality.

References

1. K. Ito, M. Mariño and H. Shu, "TBA equations and resurgent Quantum Mechanics," JHEP **01** (2019), 228 [arXiv:1811.04812 [hep-th]].
2. K. Ito and H. Shu, "TBA equations for the Schrödinger equation with a regular singularity," J. Phys. A **53** (2020) no.33, 335201 [arXiv:1910.09406 [hep-th]].

3. A. Voros, "The return of the quartic oscillator". The complex WKB method, Annales de l'I.H.P. Physique Théorique **39** (1983), no. 3 211–338.

4. Y. Emery, "TBA equations and quantization conditions," JHEP **07** (2021), 171 [arXiv:2008.13680 [hep-th]].

5. N. A. Nekrasov and S. L. Shatashvili, "Quantization of Integrable Systems and Four Dimensional Gauge Theories," [arXiv:0908.4052 [hep-th]].

6. D. Gaiotto, "Opers and TBA," [arXiv:1403.6137 [hep-th]].

7. K. Ito and H. Shu, "ODE/IM correspondence and the Argyres-Douglas theory," JHEP **08** (2017), 071 [arXiv:1707.03596 [hep-th]].

8. K. Ito, T. Kondo, K. Kuroda and H. Shu, "WKB periods for higher order ODE and TBA equations," JHEP **10** (2021), 167 [arXiv:2104.13680 [hep-th]].

9. K. Ito, T. Kondo and H. Shu, "Wall-crossing of TBA equations and WKB periods for the third order ODE," [arXiv:2111.11047 [hep-th]].

10. A. B. Zamolodchikov, "On the thermodynamic Bethe ansatz equations for reflectionless ADE scattering theories," Phys. Lett. B **253** (1991), 391-394

11. H. W. Braden, E. Corrigan, P. E. Dorey and R. Sasaki, "Affine Toda Field Theory and Exact S Matrices," Nucl. Phys. B **338** (1990), 689-746

Aspects of 5d Seiberg-Witten Theories on $\mathbb{S}^1$

Qiang Jia and Piljin Yi

Korea Institute for Advanced Study,
Seoul, Korea

In this note, we discuss the infrared physics of 5d $\mathcal{N} = 1$ Yang-Mills theories compactified on $\mathbb{S}^1$, with a view toward 4d and 5d limits. The Coulomb phase boundaries in the decompactification limit are given particular attention and related to how the wall-crossings by 5d BPS particles turn off. On the other hand, the elliptic genera of magnetic BPS strings do wall-cross and retain the memory of 4d wall-crossings, which we review with the example of dP_2 theory.

Keywords: Gauge theory; Wall-crossing.

1. Introduction

When one realizes 5d $\mathcal{N} = 1$ theories by geometric engineering as M-theory on a local Calabi-Yau[2-6], BPS objects are realized by M2 and M5 branes wrapping 2-cycles and 4-cycles respectively. The former gives electrically charged particles, including dyonic instantons, while the latter gives magnetic strings. Compared to their 4d counterpart[7,8], these 5d theories look very simple; the 5d prepotential is at most a piecewise cubic function of the real Coulombic vev's, while on $\mathbb{S}^1$, one must deal with the special Kähler geometry of complex vacuum expectation values.

The simplicity of 5d $\mathcal{N} = 1$ theories is gratifying but at times appears too simple in that the 5d theory and the same theory compactified on a circle $\mathbb{S}^1$ seem superficially very disparate. The latter acquires a much richer character. This is partly because the compactification produces particle-like monopoles from magnetic strings wrapping $\mathbb{S}^1$ and allows wall-crossing in the 4d sense. In the compactified theory, the wall-crossing involve BPS particles carrying several kinds of charges and could be very complicated. But on the other hand, in the 5d limit where $R_5 \to \infty$, the ubiquitous wall-crossing phenomena of 4d[7,9] turns off in 5d, as far as particle-like BPS states are concerned.

In this note, we wish to study 5d rank one theories compactified on a circle to study how the decompactification limit emerges in the large radius limit and how the complicated wall-crossing turns off and characterizes the co-dimension-one boundary of the Coulomb phase. We will see shortly the two are closely related to each other, and one recovers the known fact that 5d Coulomb phases end where a BPS magnetic string becomes tensionless. For some theories, this tensionless limit coincides with the symmetry restoration of the non-Abelian gauge symmetry, but not always.

On the other hand, the absence of wall-crossing really addresses the stability of 5d BPS particle states, whose central charges are all "real." 4d monopoles uplift to monopole strings, and the central charge density thereof is by and large imaginary. In fact, the Kaluza-Klein particles also come with an imaginary central charge. This explains why, upon $\mathbb{S}^1$ compactification, the wall-crossing turns on immediately. For a finite size of this circle $\mathbb{S}^1$, the magnetic strings remain extended, so the relevant counting should be given by the elliptic genus[18,19]. In the second part of this note, we will take a close look at the wall-crossing of this elliptic genus across a point where a charged matter becomes massless, or equivalently, across a flop transition.

This note is a short version of Ref[1] and is organized as follows. In Section 2, we will revisit the question of the wall-crossing starting from the circle-compactified theory and addressing what to expect from the decompactification limit. Exactly what we mean by the absence of 5d wall-crossing is mulled over, from which we draw the anticipation that the boundary of the 5d Coulomb phase should be universally characterized by tensionless BPS strings. Section 3 will take the simplest types of rank one non-Abelian theories on a circle and investigate the Coulomb phase detail. Here we will see how, in the decompactification limit, the 5d Coulomb phase boundary is realized to make sure that certain marginal walls collapse, as suggested by the general discussion on wall-crossings. The wall-crossing of magnetic BPS string across such a flop transition is the topic for Section 4, where we illustrate using the example of dP_2 near a massless quark point.

2. Wall-Crossing or Not

When it comes to BPS spectra, a very distinctive feature of the 5d theory is the absence of wall-crossing for BPS particles. Let us see how this happens and what this absence means. From the low energy dynamics of BPS objects, the usual 4d wall-crossing occurs due to runaway Coulombic directions emerging at special values of FI constants[20-22], which is in turn related to how the central charge phases are complex and phases of a pair can align at a co-dimension-one wall. With mutual intersection number, i.e., with nonzero Schwinger product of charges, this leads to a wall-crossing. On the other hand, the 5d BPS particles would be all represented by M2 wrapped on 2-cycles so that a pair of 2-cycles in a Calabi-Yau 3-fold cannot have a mutual intersection number. While D4 on a 4-cycle can have an intersection number against D2 on a 2-cycle, the former is an extended object in the form of the BPS string since it is really M5 wrapping a 4-cycle. In fact, most BPS objects that would have entered the wall-crossing quiver gauge quantum mechanics in the 4d limit are made up of magnetic strings.

Once compactified on a circle $\mathbb{S}^1$ of sufficiently small radius R_5, however, the low energy dynamics are again well captured by BPS quiver quantum mechanics for D4-D2-D0 bound states, for which the wall-crossing are numerous. We will ask exactly how the wall-crossing turns itself off back in the limit $R_5 \to \infty$. One quick answer to this is that since all magnetic objects become a string, as pointed out already, the question becomes moot if we ask for wall-crossing among 5d BPS particles. However, we can ask a little more by keeping track of BPS strings wrapped on $\mathbb{S}^1$ considered as BPS particles. As we will see below, approaching the decompactification limit from the compactified theory, the disappearance of wall-crossing occurs for multiple reasons, different for different BPS objects.

The usual 4d wall-crossing, where the complex central charge Z of the bound state aligns with Z_i and Z_j and decays as

$$Z \to Z_i + Z_j \ , \tag{1}$$

is possible only if D-branes that constitute particles i and j share a net number of open strings attached to them. This is in turn counted by the intersection number between the two cycles wrapped by the D-branes. In the $\mathbb{R}^{3+1}$ gauge theory viewpoint, on the other hand, the intersection number translates to the Schwinger product of electromagnetic charges,

$$q_i g_j - q_j g_i \neq 0, \tag{2}$$

so at least one of the two constituents must carry a magnetic charge. As such, one can imagine three logical possibilities for the 4d wall-crossing:

- Z is non-magnetic, while both constituent Z_i and Z_j are magnetic,
- Z and one of the two constituents, say Z_i, are magnetic,
- all three are magnetic.

Also, the marginal stability wall of such decay, found by equating phases of Z_i to that of Z_j, would extend between $Z_i = 0$ and $Z_j = 0$ locus in the Seiberg-Witten moduli space.

For the first case where Z_i and Z_j are magnetic, since the magnetic central charge can be approximated as:

$$Z_{\text{magnetic}} \approx iR_5 T_{\text{monopole}}, \qquad (3)$$

which is purely imaginary and T_{monopole} is the tension of the monopole string. In the 5d limit it should remain finite otherwise Z_i and Z_j will become too massive for a non-magnetic particle Z to decay to. That indicates the tension of the monopole string T_{magnetic} should go to zero at the same time when $R_5 \to \infty$. If we take the absence of such wall-crossings as given, this in turn implies that no magnetically charged BPS strings should become tensionless, perhaps except at the boundary of the Coulombic phase. This is quite natural since from the (p, q) web realization of IIB theory, where the tensionless limit of magnetic strings translates to a collapse of a face, so the above claim that the Coulomb phase ends where a magnetic BPS becomes tensionless is quite natural already.

For the second and the third, Z would represent a magnetically charged BPS string wrapped on $\mathbb{S}^1$. The second type of marginal stability wall would emanate from a point in the moduli space where some components of charged BPS particles become massless. For instance, imagine a quark hypermultiplet with mass μ_f in the defining representation of the gauge group. Since for large R_5, the central charge of Z_i is dominated by Z_{magnetic} which is purely imaginary, the wall emanating from the massless quark point following

$$\text{Arg}(Z_i Z_j^*) = 0, \qquad (4)$$

would initially extend along the circular Wilson-line direction of the small period $1/R_5$. Taking the $R_5 \to \infty$ limit, this means that there is a sense of discontinuity for magnetically charged strings across $\phi = \mu_f$, where ϕ is the 5d moduli on the Coulomb phase.

Note that $\phi > \mu_f$ is precisely where an Sp(1) doublet fermion would contribute a Jackiw-Rebbi zero-mode[26] to the monopole. With $\phi < \mu_f$, this zero-mode is lifted, so the $d = 1 + 1$ low energy dynamics of the monopole (dyon) strings change qualitatively across this point. This discontinuity translates to the conventional wall-crossing once the BPS string wraps a circle, $\mathbb{S}^1$, which may be captured via the elliptic genus of such a magnetic string; later, we will review this with a concrete case of a magnetic string in the dP$_2$ theory, or Sp(1) theory with a single massive flavor.

Marginal stability walls of the third type where magnetic strings decay into a pair of magnetic strings deserve different considerations. In 4d, such decays involving three types of dyons are known, and these walls actually extend into the weak coupling region[23]. However, for this type of wall-crossings, two independent adjoint scalar fields spanning the Coulombic moduli space are essential; The decaying states may be visualized by an analog of planar (p, q) string web, which cannot be drawn when the adjoint scalar is real[24]. Such walls would not survive the decompactification limit since half of the Coulombic directions, corresponding to the Wilson line vev, collapses due to $R_5 \to \infty$. Thus, such 3rd type of marginal stability would also turn off in the decompactification limit, leaving behind the simplest possible chamber.

3. Sp(1) Theories on a Large $\mathbb{S}^1$

In this section we will analyse the pure 5d Sp(1) theory on a large $\mathbb{S}^1$ and illustrate how does the second type of marginal stability walls behave in the strict 5d limit. In particular, we will restrict to the Sp(1) theory with zero theta angle, namely the F0 theory, for a general discussion involving Sp(1) theory please refer to Ref[1].

The Coulomb phase of 5d $\mathcal{N} = 1$ supersymmetric gauge theory is famously described by the Intriligator-Morrison-Seiberg (IMS) prepotential[5], which is one-loop exact. In particular, the prepotential for Sp(1) theory is:

$$\mathcal{F}_{\text{IMS}}(\phi) = \mu_0 \phi^2 + \frac{4}{3}\phi^3 \, , \tag{5}$$

where ϕ is the real Coulmob phase moduli and $\mu_0 \equiv 8\pi^2/g_5^2$ is the 5d inverse coupling-squared and also the instanton mass. The first derivative gives the monopole string tension T_{mono} as,

$$iT_{\text{mono}} = \frac{i}{2\pi}\frac{\partial \mathcal{F}_{\text{IMS}}}{\partial \phi} = \frac{i2\phi(\mu_0 + 2\phi)}{2\pi}, \tag{6}$$

where we used iT_{mono} on the left hand side as a reminder that the 4d monopole central charge upon a compactification is imaginary in the large radius limit. The second derivative produces the pure imaginary 5d coupling as

$$\tau_{\text{5d}} = \frac{i}{2\pi}\frac{\partial^2 \mathcal{F}_{\text{IMS}}}{\partial \phi^2} = \frac{i(\mu_0 + 4\phi)}{\pi}, \tag{7}$$

with $\tau_{\text{5d}} = i8\pi/g_{5,\text{eff}}^2$.

58

After compactification on $\mathbb{S}^1$, the prepotential will receive instanton corrections, which becomes [6,11-14]:

$$\frac{\partial^3 \mathcal{F}_{5d,\mathbb{S}^1}(a,\mu_0)}{\partial a^3}$$
$$= 8 + \sum_{m,n \geq 0} N_{m,n} \left(\frac{e^{-4\pi m a R_5} e^{-2\pi n (2a+\mu_0) R_5}}{1 - e^{-4\pi m a R_5} e^{-2\pi n (2a+\mu_0) R_5}} \right) (-2m - 2n)^3, \quad (8)$$

where a is the complexified version of the 5d moduli ϕ combining with the Wilson line around the circle, and in the asymptotic region of the Coulombic moduli space one has $\mathrm{Re}(a) \approx \phi$. $N_{m,n}$ is an invariant which counts the number of the holomorphic 2-cycles in Calabi-Yau picture, where the exact numbers can be found in Ref. [14]. We are mostly interested in the interface between 4d and 5d, so we take R_5 to be sufficiently large in the following discussion whereby $|g_5^2/2\pi R_5| \ll 1$.

Let's assume the bare coupling square is positive in the following discussion. Since we are working with large R_5 ($\mu_0 R_5 \gg 1$), one can set $n = 0$ in (8) to obtain:

$$\frac{\partial^3 \mathcal{F}_{5d,\mathbb{S}^1}}{\partial a^3} = 8 + 16 \left(\frac{e^{-4\pi a R_5}}{1 - e^{-4\pi a R_5}} \right), \quad (9)$$

where we use the fact that the only non-zero $N_{m,0}$ is $N_{1,0} = -2$ for F0 [14]. The effective 4d coupling for the compactified theory is obtained by integration:

$$\tau_{4d} = 2\pi R_5 \tau_{5d} \approx i\frac{2}{\pi} \log \left(\frac{4 \sinh^2(2\pi a R_5)}{(2\pi R_5 \Lambda_{\mathrm{QCD}})^2} \right), \quad (10)$$

where $\tau_{5d} = \frac{i}{2\pi} \frac{\partial^2 \mathcal{F}_{5d,\mathbb{S}^1}}{\partial \phi^2}$ is defined parallel to (7) and the integration constant is determined by comparing (7). Here Λ_{QCD} is the effective QCD scale for the compactified theory defined as $2\pi R_5 \Lambda_{\mathrm{QCD}} \equiv e^{-\pi \mu_0 R_5/2}$, such that if $\mathrm{Re}(a) \ll 1/2\pi R_5$ and the 4d approximation can be trusted, (10) indicates:

$$\frac{1}{g_{4,\mathrm{eff}}^2} \approx \frac{1}{4\pi^2} \log \left(\frac{(2a)^2}{\Lambda_{\mathrm{QCD}}^2} \right), \quad (11)$$

which is the correct 4d coupling implied by β-function.

The moduli space for F0 theory with positive μ_0 is depicted as figure 1[a]. There is a strongly coupled region $\phi \sim \Lambda_{\mathrm{QCD}}$ near the tip of the cigar,

---

[a]Actually for a generic Sp(1) theory with matters, the moduli space is a double-copy of figure.1 and there are two distinct holonomy saddles. We refer Ref. [15] for a detailed discussion of the moduli space.

and the theory is effectively the strongly coupled pure Sp(1) Seiberg-Witten theory. Around Λ_{QCD} there are two singularities where the mass of $(0,1)$ monopole and $(2,-1)$ dyon will separately become zero, and there is a marginal stability wall connecting these two singularities as shown in the figure 1.

Now let's see what happens in the decompactification limit $R_5 \to \infty$. Since the instanton contribution vanishes in this limit, we can identify $\text{Re}(a) = \phi$. The period of a is $a \sim a + i/2R_5$, therefore as $R_5 \to \infty$ the imaginary part of a is zero such that $a = \phi$ in this limit, and the cylinder will become the half-line in figure 1. Also Λ_{QCD} becomes much smaller than $1/2\pi R_5$ in this limit therefore the two singularities and the marginal stability wall both shrink to the endpoint $\phi = 0$ in the 5d moduli space. At that point both W-bosons and monopoles become massless and the Sp(1) gauge symmetry is restored.

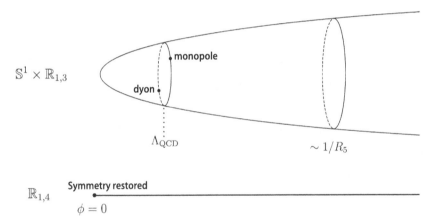

Fig. 1. Moduli space for F0-theory with positive coupling-squared. In the large R_5 limit, entire strong coupling region is pushed to the far left of $1/R_5$, which itself is pushed toward $\phi = 0$. No trace of the marginal stability wall is left behind.

We have discussed the first kind of marginal stability wall where the W-boson decays into a pair of $(0,1)$ monopole and $(2,-1)$ dyon and seen how the wall collapses to the endpoint of the 5d moduli space. More generally, for any pair of magnetic particles Z_i and Z_j, their central charge are of order $\sim \Lambda_{\text{QCD}}$ in the core region or dominated by the magnetic part:

$$Z_{\text{mag}} \sim iR_5 2a(2a + \mu_0), \tag{12}$$

in the asymptotic region $\text{Re}(a) > 1/2\pi R_5$.

On the other hand, since Z is non-magnetic, its central charge is a combination of the electric, instanton and KK parts:

$$Z_{\text{elec}} = 2a, \quad Z_{\text{Inst}} = \mu_0, \quad Z_{\text{KK}} = \frac{i}{R_5}. \tag{13}$$

The wall, if any, must remain in the region where $Z_{\text{elec}}, Z_{\text{Inst}}, Z_{\text{KK}}$ are comparable with Z_{mag}, end where the magnetic particles Z_i or Z_j become massless. $Z_{\text{elec}} \sim Z_{\text{mag}}$ requires $a \sim \Lambda_{\text{QCD}}$ while $Z_{\text{Inst}} \sim Z_{\text{mag}}$ can happen only if $\text{Re}(a) \sim 1/2\pi R_5$. Z_{KK} scales down as $1/R_5$ also. Therefore, no walls of the first kind can extend beyond $a \sim 1/2\pi R_5$, which in turn collapse to $\phi = 0$ in the decompactification limit.

4. Wall-Crossing of a Magnetic BPS String

In this section we will explore the second type of marginal stability wall, where the bound state Z and one of its component Z_i are magnetic. As discussed in the previous section, one may consider adding a flavour with positive mass μ_f to the 5d theory, and on the Coulomb branch it becomes massless at $\phi = \mu_f$. From the M-theory picture, the quark is represented by a M2 brane wrapping on a compact 2-cycle, and at the massless point in the Coulomb branch the volume of the 2-cycle vanishes, which is responsible for a flop transition of the local Calabi-Yau.

The correct description of these magnetic objects is, in fact, well known to be governed by $(0, 4)$ nonlinear sigma models which has been well-studied in various literatures[16,17]. The elliptic genus that counts the BPS states for such a theory is defined as[18,19]:

$$Z(\tau, \bar{\tau}, y) = \text{Tr}_R \frac{1}{2} F^2 (-1)^F e^{\pi i D_{ab} s^a q^b} q^{L_0 - \frac{c_L}{24}} \bar{q}^{\bar{L}_0 - \frac{c_R}{24}} e^{2\pi i y_a q^a}, \tag{14}$$

where the monopole string is represented by the M5-brane wrapping on the compact divisor S in the local Calabi-Yau $\hat{X}$, q^a are membrane charges, D_{ab} is the intersection matrix between 2/4-cycles and $s^a/2$ is a half-integer charge shift due to the Freed-Witten anomaly[25].

The evaluation of the elliptic genus is based on a fact that the elliptic genus $Z(\tau, \bar{\tau}, y)$ is subject to a θ-function decomposition (see[1,20]for a review). Eventually the elliptic genus is totally determined by the geometry of the compact divisor S and the way it is embedded in the ambient Calabi-Yau $\hat{X}$. The change of geometry during flop transition will induce a jump of the elliptic genus, and for dP$_2$ geometry that is:

$$Z_{\text{dP}_2} = -Z_{\widetilde{\text{dP}}_2} \times \frac{\theta_{11}(\tau, y_3)}{\eta(\tau)}, \tag{15}$$

where the θ-part is exactly the partition function of a chiral fermion generated by the Jackiw-Rebbi zero-mode as discussed in section 2.

References

1. Q. Jia and P. Yi, [arXiv:2111.09448 [hep-th]].
2. N. Seiberg, "Five-dimensional SUSY field theories, nontrivial fixed points and string dynamics," Phys. Lett. B **388**, 753-760 (1996) [arXiv:hep-th/9608111 [hep-th]].
3. D. R. Morrison and N. Seiberg, "Extremal transitions and five-dimensional supersymmetric field theories," Nucl. Phys. B **483**, 229-247 (1997) [arXiv:hep-th/9609070 [hep-th]].
4. M. R. Douglas, S. H. Katz and C. Vafa, "Small instantons, Del Pezzo surfaces and type I-prime theory," Nucl. Phys. B **497**, 155-172 (1997) [arXiv:hep-th/9609071 [hep-th]].
5. K. A. Intriligator, D. R. Morrison and N. Seiberg, "Five-dimensional supersymmetric gauge theories and degenerations of Calabi-Yau spaces," Nucl. Phys. B **497** (1997), 56-100 [arXiv:hep-th/9702198 [hep-th]].
6. S. H. Katz, A. Klemm and C. Vafa, "Geometric engineering of quantum field theories," Nucl. Phys. B **497** (1997), 173-195 [arXiv:hep-th/9609239 [hep-th]].
7. N. Seiberg and E. Witten, "Electric - magnetic duality, monopole condensation, and confinement in N=2 supersymmetric Yang-Mills theory," Nucl. Phys. B **426**, 19-52 (1994) [erratum: Nucl. Phys. B **430**, 485-486 (1994)] [arXiv:hep-th/9407087 [hep-th]].
8. N. Seiberg and E. Witten, "Monopoles, duality and chiral symmetry breaking in N=2 supersymmetric QCD," Nucl. Phys. B **431**, 484-550 (1994) [arXiv:hep-th/9408099 [hep-th]].
9. F. Ferrari and A. Bilal, "The Strong coupling spectrum of the Seiberg-Witten theory," Nucl. Phys. B **469**, 387-402 (1996) [arXiv:hep-th/9602082 [hep-th]].
10. Z. Duan, D. Ghim and P. Yi, "5D BPS Quivers and KK Towers," JHEP **02**, 119 (2021) [arXiv:2011.04661 [hep-th]].
11. P. Candelas, X. C. De La Ossa, P. S. Green and L. Parkes, "A Pair of Calabi-Yau manifolds as an exactly soluble superconformal theory," Nucl. Phys. B **359**, 21-74 (1991)
12. S. H. Katz, D. R. Morrison and M. R. Plesser, "Enhanced gauge symmetry in type II string theory," Nucl. Phys. B **477** (1996), 105-140 [arXiv:hep-th/9601108 [hep-th]].

13. A. E. Lawrence and N. Nekrasov, "Instanton sums and five-dimensional gauge theories," Nucl. Phys. B **513** (1998), 239-265 [arXiv:hep-th/9706025 [hep-th]].

14. T. M. Chiang, A. Klemm, S. T. Yau and E. Zaslow, "Local mirror symmetry: Calculations and interpretations," Adv. Theor. Math. Phys. **3** (1999), 495-565 [arXiv:hep-th/9903053 [hep-th]].

15. C. Closset and H. Magureanu, [arXiv:2107.03509 [hep-th]].

16. J. M. Maldacena, A. Strominger and E. Witten, "Black hole entropy in M theory," JHEP **12** (1997), 002 [arXiv:hep-th/9711053 [hep-th]].

17. R. Minasian, G. W. Moore and D. Tsimpis, "Calabi-Yau black holes and (0,4) sigma models," Commun. Math. Phys. **209** (2000), 325-352 [arXiv:hep-th/9904217 [hep-th]].

18. J. de Boer, M. C. N. Cheng, R. Dijkgraaf, J. Manschot and E. Verlinde, "A Farey Tail for Attractor Black Holes," JHEP **11**, 024 (2006) [arXiv:hep-th/0608059 [hep-th]].

19. D. Gaiotto, A. Strominger and X. Yin, "The M5-Brane Elliptic Genus: Modularity and BPS States," JHEP **08**, 070 (2007) [arXiv:hep-th/0607010 [hep-th]].

20. F. Denef and G. W. Moore, "Split states, entropy enigmas, holes and halos," JHEP **11**, 129 (2011) [arXiv:hep-th/0702146 [hep-th]].

21. F. Denef, "Quantum quivers and Hall / hole halos," JHEP **10**, 023 (2002) [arXiv:hep-th/0206072 [hep-th]].

22. K. Hori, H. Kim and P. Yi, "Witten Index and Wall Crossing," JHEP **01**, 124 (2015) [arXiv:1407.2567 [hep-th]].

23. K. M. Lee and P. Yi, "Dyons in N=4 supersymmetric theories and three pronged strings," Phys. Rev. D **58**, 066005 (1998) [arXiv:hep-th/9804174 [hep-th]].

24. O. Bergman and B. Kol, "String webs and 1/4 BPS monopoles," Nucl. Phys. B **536**, 149-174 (1998) [arXiv:hep-th/9804160 [hep-th]].

25. D. S. Freed and E. Witten, "Anomalies in string theory with D-branes," Asian J. Math. **3**, 819 (1999) [arXiv:hep-th/9907189 [hep-th]].

26. R. Jackiw and C. Rebbi, "Solitons with Fermion Number 1/2," Phys. Rev. D **13**, 3398-3409 (1976)

de Sitter Duality and Holographic Renormalization

Yoshihisa KITAZAWA[1),2), a]

[1)] *KEK Theory Center, Tsukuba, Ibaraki 305-0801, Japan*
[2)] *Department of Particle and Nuclear Physics*
The Graduate University for Advanced Studies (Sokendai)
Tsukuba, Ibaraki 305-0801, Japan

We perform the resummation of the infrared logarithms in the inflationary universe. Applying the renormalization group, we derive the stochastic equations as the effective theory at the horizon. We focus on the conformal zero mode to respect local Lorentz symmetry. Under Gaussian approximation, we derive the fundamental equation for the Universe (EqU). We also derive the identical equation from the first law of thermodynamics in a dual geometric picture. We believe it is a convincing evidence for de Sitter duality between quantum stochastic physics on the boundary and classical thermodynamics in the bulk. The equation for the Universe (EqU) possesses the solution with the ultraviolet fixed point. It also contains the inflationary universe with the power potentials. We discuss possible scenarios for the very early universe with decreasing ϵ. We argue inflationary universe subsequently dominates to maximize the entropy and ϵ problem is naturally solved.

1. Introduction

dS space may be decomposed into the bulk and the boundary, i.e., the sub-horizon and horizon. From a holographic perspective, we consider the conformal zero mode dependence of the Einstein-Hilbert action:

$$\frac{1}{16\pi G_N} \int \sqrt{g} d^4 x (Re^{2\omega} - 6H^2 e^{4\omega}) \simeq \frac{\pi}{G_N H^2}(1 - 4\omega^2), \qquad (1)$$

where the gauge fixing sector is suppressed because it does not produce the IR logarithms in the background gauge. Our gauge fixing procedure and the propagators are explained in [4]. The semi-classical dS entropy was obtained by rotating the background spacetime dS^4 to S^4 in (1)[5].

---

[a]E-mail address: kitazawa@post.kek.jp

The quadratic part of ω constitutes a Gaussian distribution function for the conformal zero mode,

$$\rho(\omega) = \sqrt{\frac{4}{\pi g}} \exp\left(-\frac{4}{g}\omega^2 \right).$$ (2)

It may represent an initial state of the Universe when the dS expansion begins. In order to describe the time evolution of the conformal factor of the Universe, we introduce a new parameter $\xi(t)$.

$$\rho(\xi(t), \omega) = \sqrt{\frac{4\xi(t)}{\pi g}} \exp\left(-\frac{4\xi(t)}{g}\omega^2 \right).$$ (3)

ξ is the only parameter in the Gaussian approximation. We work within the Gaussian approximation since it is an excellent approximation for gravity with the small coupling g. The von Neumann entropy $S = -\mathrm{tr}(\rho \log \rho) \sim 1/2 \log(g/\xi)$ becomes larger as ξ becomes smaller. Thus the diffusion triggers an instability in de Sitter space.

In terms of the distribution function, the n-point functions are defined as follows

$$\langle \omega^n(t) \rangle_{\mathrm{boundary}} = \int d\omega \rho(\xi(t), \omega)\omega^n.$$ (4)

In particular, the two-point function of the conformal mode is given by

$$\langle \omega^2(t) \rangle_{\mathrm{boundary}} = \frac{g}{8\xi(t)}.$$ (5)

The negative norm of the bulk conformal mode indicates that the ρ diffuses toward the future. In fact the perturbative quantum expectation of (1) gives $g(t) \sim g(1 - 2\gamma Ht) \sim g(1 - 3gHt)$. The duality between the inflation and quantum gravity is also based on this one loop effect[6,7]. However such an estimate is reliable only locally $Ht \ll 1$. In cosmology it is essential to resum all powers of IR logarithms $(Ht)^n$ to understand the global picture. For such a purpose, we find our holographic approach is up to the task.

We investigate the dynamics of conformal mode after integrating the bulk mode. The two point function gives rise to IR logarithms.

$$< \omega^2 >_{\mathrm{bulk}} = -\frac{3g}{4} \int_{Ha(t)}^{\Lambda} \frac{dk}{k}$$

$$= \frac{3g}{4} Ht = \frac{3g}{4} N(t).$$ (6)

We assume there is no time dependent UV contributions. We focus on the Hubble scale physics where $a(t) = 1/ - \tau H = \exp(Ht)$. We recall that $-\tau\mu \sim 1$ at the Horizon. While the wave functions of the bulk mode oscillates, those of the boundary mode do not. That is why we call it the zero mode. Our strategy is to integrate out oscillating mode first.

The finite bare distribution function is given by subtracting the bulk mode contribution above the Hubble scale. We thus construct low energy effective theory around the Hubble scale. Such a theory is holographic and finite after the subtraction.

$$\rho_B = \exp\left(-\frac{3g}{4}\frac{Ht}{2}\frac{\partial^2}{\partial\omega^2}\right)\rho. \tag{7}$$

The renormalization scale of the low energy effective action is the Hubble scale. As ρ_B is independent of the renormalization scale, the renormalized distribution function obeys the following renormalization group equation.

$$\dot{\rho} - \frac{3g}{4}\cdot\frac{H}{2}\frac{\partial^2}{\partial\omega^2}\rho = 0, \tag{8}$$

where $\dot{O}$ denotes a derivative of O with respect to the cosmic time t. The factor $3g/4$ in the diffusion term is the projection factor to the IR region. The conformal mode ω consists of a minimally coupled field X. $\omega = \sqrt{3}X/4 + Y/4$[4,8]. We neglect Y because it has the effective mass $m_{eff}^2 = 2H^2$. There is no drift term in the reduced space consisting only of X. In fact X is nothing other than the curvature perturbation ζ.

The gravitational FP equation (8) is obtained by integrating the quantum bulk modes inside the horizon. It turns out to be a diffusion equation due to the lack of the drift term. The solution is the Brownian motion as it is jolted by the horizon exiting modes. The FP equation is a dynamical renormalization group equation. We can sum up the IR logarithms $\log^n a = (Ht)^n$ by this equation to find a running coupling $g(t)$.

The FP (diffusion) equation shows that the solution is the Gaussian distribution with the standard deviation increasing linearly with the e-folding number $N(t)$[9].

$$\frac{3}{4}gHt = \frac{3g}{4}N(t) = \frac{g}{8\xi}. \tag{9}$$

The standard deviation is related to N as $1/\xi = 6N$ in (9). It is consistent with a standard Brownian motion prediction. The von Neumann entropy thus increases logarithmically,

$$\delta S = \frac{1}{2} \log \frac{1}{\xi} = \frac{1}{2} \log 6N(t). \tag{10}$$

Identifying the von Neumann entropy of conformal zero mode with the quantum correction to dS entropy, we obtain the bare action with the counter term

$$\frac{1}{g_B} = \frac{1}{g(N)} - \frac{1}{2} \log(6N). \tag{11}$$

By requiring the bare action is independent of the renormalization scale: namely N, we obtain the one loop β function.

$$\beta = \frac{\partial}{\partial \log(N)} g(N) = -\frac{1}{2} g(N)^2. \tag{12}$$

We find the running gravitational coupling as

$$g(N) = \frac{2}{\log(N)}. \tag{13}$$

The holographic investigation at the boundary shows that g is asymptotically free toward the future[10,11]. The renormalization group trajectory must reach Einstein gravity in the weak coupling limit for the consistency with general covariance[18]. We find that it approaches a flat spacetime in agreement with this requirement.

2. Equation for the Universe and de Sitter Duality

The Gaussian distribution of the conformal zero mode is characterized by the standard deviation $1/\xi$. Although there is no inflaton in Einstein gravity, we propose to identify the inflaton f^2 as $f^2 \propto 1/\xi$. In our interpretation, the inflaton is not a fundamental field but a stochastic variable. The two point function at an equal time grows due to the Brownian motion: IR logarithmic fluctuations $1/\xi \sim N$. While the inflation theory is specified by the inflaton potential, the dynamics of quantum gravity is determined by the FP equation which describes the stochastic process at the horizon. We thus argue the classical solution of the inflation theory satisfies the FP equation as well.

It is likely that there are multiple elements in the universality class of quantum gravity/inflation theory. The inflation era of the early Universe may be one of them. As we discuss shortly we find a pre-inflation era which is indispensable to launch inflation era which in turn necessary to trigger the big bang.

We evaluated the time evolution of entropy to the leading log order in (10). In order to take account of the higher loop corrections in g, the FP equation should be generalized. It turns out to be just necessary to introduce the dimensionless gravitational coupling $g(t)$ and e-folding number: $N = \int^t dt' H(t')$.

$$\frac{\partial}{\partial N} \rho_N - \frac{3g(t)}{4} \cdot \frac{1}{2} \frac{\partial^2}{\partial \omega^2} \rho_N = 0. \tag{14}$$

Since the equation is local, we consider the following local solutions.

$$\rho = \sqrt{\frac{4\xi(t)}{\pi g(t)}} \exp\left(-\frac{4\xi(t)}{g(t)}\omega^2\right). \tag{15}$$

We next put the ansatz into the FP equation and find the condition for the background to satisfy.

$$\frac{\partial}{\partial N} \frac{g(t)}{8\xi} = \frac{3}{4}g. \tag{16}$$

We obtain a remarkably simple equation.

$$\frac{\partial}{\partial N} \log \frac{g(t)}{\xi} = 6\xi. \tag{17}$$

(17) determines the evolution of von Neumann entropy $S = \frac{1}{2}\log\frac{g}{\xi}$ with respect to N. This formula shows the validity of our postulate that von Neumann entropy of conformal zero mode constitutes the quantum correction to de Sitter entropy. We call it the equation for the Universe (EqU).

The increase of the entropy $S = 1/g$ can be evaluated by the first law $T\Delta S = \Delta E$ where ΔE is the incoming energy flux of the inflaton[22]. In this way, the one of the Einstein's equation is obtained:

$$\dot{H}(t) = -4\pi G_N \dot{f}^2. \tag{18}$$

From this formula, we obtain

$$2\epsilon = -\frac{\partial}{\partial N} \log g(t). \tag{19}$$

We add the same quantity $1/\tilde{N}$ to the both sides of the equation.

$$2\epsilon + \frac{1}{\tilde{N}} = -\frac{\partial}{\partial N} \log(g\tilde{N}). \tag{20}$$

For power potential inflationary universe, (20) is rewritten as

$$6\xi = \frac{\partial}{\partial N} \log \frac{g}{\xi}.$$

It is precisely our EqU (17). We have derived it in a dual geometric picture here in place of the original quantum stochastic picture.

The equation (17) has inflationary solutions with power potentials.

$$g = c\tilde{N}^{\frac{m}{2}}, \quad \xi = \frac{m+2}{12\tilde{N}}. \tag{21}$$

Here c is an integration constant. m denotes the power of the potential: f^m. It is convenient to replace N by $\tilde{N}$ where $\tilde{N} = N_e - N$. N_e denotes the e-foldings at the end of inflation. (17) becomes as follows after the change of the variable,

$$-\frac{\partial}{\partial N} \log \frac{g(\tilde{N})}{\xi(\tilde{N})} = 6\xi(\tilde{N}). \tag{22}$$

Although the dS entropy can be explained by quantum effects alone for the weakly coupled inflaton solution, the strongly coupled inflaton solution is a dual object in the sense that geometrical description is reliable.

It proves the consistency of the EqU and the confirmation of de Sitter duality. Since (17) and (19) are equivalent, the solutions of the former satisfy the latter.

3. Inflation and UV completion

In the literature, δN formalism is widely used to investigate the curvature perturbation. It underscores the validity of the stochastic picture of the inflation.[12-16] Let us consider the fluctuation of the curvature perturbation ζ.

$$\zeta = \delta N = \frac{H}{\dot{f}}\delta f,$$

$$< \delta f(t)\delta f(t') > = (\frac{H^2}{4\pi^2})H\delta(t - t'). \tag{23}$$

We obtain in the super-horizon regime:

$$\frac{\partial}{\partial N} < \zeta^2(t) > = \frac{\partial}{\partial N} < (\frac{H}{\dot{f}})^2\delta f^2 >$$

$$= \frac{1}{2\epsilon M_P^2}\frac{\partial}{\partial N} < \delta f^2 >$$

$$= \frac{H^2}{8\pi^2\epsilon M_P^2} = \frac{g}{\epsilon(t)} = P. \tag{24}$$

We recall the following identity holds at the horizon exit $t = t_*$

$$\dot{\rho}(t_*)e^{\rho(t_*)} = k. \tag{25}$$

It is nothing but choosing our renormalization scale as $\log k = Ht$. Let $P \sim k^{n_s - 1}$. The scaling dimension of P can be estimate as

$$k \frac{d}{dk}(4 \log \dot{\rho}(t_*) - 2 \log \dot{\phi}(t_*))$$
$$= \frac{d}{dN_*}(4 \log \dot{\rho}(t_*) - 2 \log \dot{\phi}(t_*))$$
$$= 2(\eta - 3\epsilon), \tag{26}$$

where

$$\epsilon = \frac{\dot{\phi}^2}{2H^2}, \quad \eta = \frac{\ddot{\phi}}{\dot{\rho}\dot{\phi}} + \epsilon. \tag{27}$$

In power inflation potential models $V \sim f^m$, ξ and ϵ scale as $1/N$ due to the universality of the random walk. They belong to the same universality class as the following scalar two point functions scale as

$$\frac{\partial}{\partial N} < \zeta^2 > = \frac{g}{\epsilon}, \quad \epsilon = \frac{m}{4N(t)}. \tag{28}$$

The scaling exponents agree with our FP equation and the δ N formalism.

$$1 - n_s = \frac{\partial}{\partial N} \log(gN(t)) = 2\epsilon + \frac{1}{N(t)} = \frac{m+2}{2N(t)}. \tag{29}$$

There is a UV fixed point in our renormalization group. We study it next. FP equation (17) enables us to evaluate higher order corrections to the β function. The expansion parameter is $1/\log N$. We can confirm that the following g_f and ξ_f satisfies (17),

$$g_f = \frac{2}{\log N}\left(1 - \frac{1}{\log N}\right), \quad \xi_f = \frac{1}{6N}\left(1 - \frac{1}{\log N}\right). \tag{30}$$

Thus, the β function, ϵ and the semi-classical entropy generation rate are given by

$$\beta = \frac{\partial}{\partial \log N} g_f = -\frac{2}{\log^2 N} + \frac{4}{\log(N)^3} = -2\left(1 - 2\frac{1}{\log N}\right)\frac{1}{\log^2 N}. \tag{31}$$

$$\epsilon_f = -\frac{1}{2}\frac{\partial}{\partial N} \log(g_f) = -\frac{1}{2g_f N}\beta_f. \tag{32}$$

$$\frac{\partial}{\partial N} S_{sc} = \frac{\partial}{\partial N}\frac{1}{g_f} = -\frac{1}{Ng_f^2}\beta_f. \tag{33}$$

A remarkable feature is that the coupling has the maximum value $g = 1/2$ at the beginning. It steadily decreases toward the future as the β

function is negative in the whole region of time flow. It has two fixed points at the beginning and at the future of the Universe. The existence of the UV fixed point may indicate the consistency of quantum gravity. The single stone solves the ϵ problem[17] as well since it vanishes at the fixed point. The β function describes a scenario that our Universe started the dS expansion with a minimal entropy $S = 2$ while it has $S = 10^{120}$ now. It corresponds to $N = e^2 \sim 7.4, a = e^N \sim 1.6 \times 10^3$. The just born Universe is rather large which reflects the critical coupling $g = 1/2$ is rather small. In terms of the reduced Planck mass, $H^2/4\pi^2 M_P^2 = 1$.

Since we work with the Gaussian approximation, our results on the UV fixed points are not water tight as the coupling is not weak. Nevertheless we find it remarkable that they support the idea that quantum gravity has a UV fixed point with a finite coupling. In fact 4 dimensional de Sitter space is constructed in the target space at the UV fixed point of $2+\epsilon$ dimensional quantum gravity[18]. 4 dimensional de Sitter space also appears at the UV fixed point of the exact renormalization group[19][20].

Such a theory might be a strongly interacting conformal field theory. However, it is not an ordinary field theory as the Hubble scale is Planck scale. Our dynamical β function is closely related to the cosmological horizon and physics around it. The existence of the UV fixed point could solve the trans-Planckian physics problem. A consistent quantum gravity theory can be constructed under the assumption that there are no degrees of freedom at trans-Planckian physics[21]. In this sense, it is consistent with string theory and matrix models. The Universe might be governed by (30) in the beginning as it might be indispensable to construct the UV finite solutions of the FP equation.

4. The Inflationary Universe and Pre-History

The inflaton may be identified with the stochastic variable f whose correlators show characteristic features of Brownian motion. $< f^2 > = \tilde{N}$ and $g \propto \tilde{N}^{m/2} = < f^m >$. $6\xi = 1 - n_s$ measures the extra tilt of the scalar two point function $k^{n_s - 1}$ with respect to k.

The left-hand side of (17) can be identified with $1 - n_s$ where n_s is the scalar spectral index. Let us recall that $1 - n_s$ is expressed by the slow-roll parameters ϵ and η,

$$1 - n_s = 6\epsilon - 2\eta. \tag{34}$$

The equation (17) is equivalent to (34) for the f^m inflaton potential where $\eta = (m-1)/(2\tilde{N})$.

We thus conclude:

$$g = c\tilde{N}^{\frac{m}{2}}, \quad \epsilon = -\frac{1}{2}\frac{\partial}{\partial N}log(g) = \frac{m}{4\tilde{N}},$$
$$1 - n_s = 6\xi = \frac{m+2}{2\tilde{N}}. \tag{35}$$

In this case, the concave power solutions can be obtained by formally replacing m by $1/n$. We admit it is not entirely clear why concave potentials are relevant. It is possible that the convex potentials are already excluded by observations.

In contrast, there is more room for concave potentials to accommodate the observational information with the judicious choice of n.

$$\epsilon = -\frac{1}{2}\frac{\partial}{\partial N}log(g) = \frac{1}{4n\tilde{N}},$$
$$1 - n_s = \frac{\frac{1}{n}+2}{2\tilde{N}}. \tag{36}$$

In our view the inflaton behaves as an particle in the Brownian motion. Its trajectory is made of infra-red fluctuations. Nevertheless the convex and the concave potentials in the inflation models seems to form two distinct groups. The concave potentials are promising avenue to explore right now[23].

Table 1.

ϵ	$n = 1$	$n = 2$	$n = 3$
	$\frac{1}{4\tilde{N}}$	$\frac{1}{8\tilde{N}}$	$\frac{1}{12\tilde{N}}$
$1 - n_s$	$\frac{3}{2\tilde{N}}$	$\frac{5}{4\tilde{N}}$	$\frac{7}{6\tilde{N}}$

In Table 1, we list expected ϵ and $1 - n_s$ in the power potential model with $f^{1/n}, n = 1, 2, 3$. We note $1 - n_s$ is bounded from below by $1/\tilde{N}$ while $r = 16\epsilon$ is not. It is consistent with the current observations. It is important to establish the bound on n. We argue the order of magnitude estimate can be trusted for the concave potentials as they are in the weak coupling regime. The semi-classical de Sitter entropy favors smaller n as the corresponding entropy grows faster: $S \sim (1/\tilde{N})^{\frac{1}{2n}}$.

Although it is still highly speculative, we argue that our Universe is likely to be located close to the point $m = n = 1$. Since the convex potentials are excluded by observation, it must be a concave potential with a small n. The convex potentials correspond to strongly coupled systems.

Here the naturalness and the anthropic principle may come in. It is hard to believe that our Universe comes out of a strongly coupled system. It is much easier to accommodate hierarchies in the weakly coupled system. Our Universe might to be located near the boundary between the stable and unstable universes like the standard model of particle physics. Since the Universe can stay at the fixed point forever, it is likely to be not far from it.

Our FP equation is expected to go beyond the resummation of leading IR logarithms which results in the logarithmic decay of g. It is still a great surprise to find inflation theory as its solutions. Fortunately EqU can be derived in a dual geometrical picture. This fact gives a non-perturbative evidence for the dS duality.

The solution (30) is UV complete. g is attracted to the fixed point as we roll back the history of the Universe. However, it cannot terminate the eternal inflation as $\epsilon = -(1/2)\partial \log g/\partial N \sim 1/(2N \log N)$ decreases with time. On the other hand, the solution (21) is not UV complete but it can end the inflation as $\epsilon \sim m/4\tilde{N}$ increases with time. These solutions generate the entropy in different ways: We consider the leading de Sitter entropy $1/g(t)$. The t dependence of $g(t)$ depends on the quantum corrections through the EqU equation. $1/g \sim \log N$ for the former and $1/g \sim 1/\tilde{N}^{\frac{1}{2n}}$ for the latter. From the perspective of the dominant entropy principle, (30) is chosen initially and (21) is chosen after $\log N \sim 1/\tilde{N}^{\frac{1}{2n}}$. That is to say, g_1 in (30) describes the newly born Universe and g_2 in (21) describes the inflation era.

References

1. A. A. Starobinsky, Lect. Notes Phys. **246**, 107 (1986); A. A. Starobinsky and J. Yokoyama, Phys. Rev. D **50**, 6357 (1994).
2. N. C. Tsamis and R. P. Woodard, Nucl. Phys. B **724**, 295 (2005).
3. H. Kitamoto and Y. Kitazawa, Int. J. Mod. Phys. A **29**, 1430016 (2014).
4. H. Kitamoto and Y. Kitazawa, Phys. Rev. D **87**, 124007 (2013).
5. G. W. Gibbons and S. W. Hawking, Phys. Rev. D **15**, 2752 (1977).
6. H. Kitamoto and Y. Kitazawa, Phys. Rev. D **99**, 085015 (2019).
7. H. Kitamoto, Y. Kitazawa and T. Matsubara, Phys. Rev. D **101**, 023504 .
8. N. C. Tsamis and R. P. Woodard, Commun. Math. Phys. **162**, 217 (1994).

9. G. Parisi, Statistical Field Theory, Frontiers in Physics **66**, (1987).)
10. D. J. Gross and F. Wilczek, Phys. Rev. Lett. **30**, 1343 (1973).
11. H. D. Politzer, Phys. Rev. Lett. **30**, 1346 (1973).
12. A. A. Starobinsky, Phys. Lett. B117, 175 (1982); JETP Lett. 42, 152 (1985).
13. D. S. Salopek and J. R. Bond, Phys. Rev. D42, 3936 (1990).
14. M. Sasaki and E. D. Stewart, Prog. Theor. Phys. 95, 71 (1996), arXiv:astro-ph/9507001.
15. D. H. Lyth, K. A. Malik, and M. Sasaki, JCAP 0505, 004 (2005), arXiv:astro-ph/0411220v3.
16. D. H. Lyth and Y. Rodriguez, Phys.Rev.Lett. 95, 121302 (2005), arXiv:astro-ph/0504045 [astro-ph].
17. R. Penrose, Annals N. Y. Acad. Sci. **571**, 249 (1989).)
18. H. Kawai, Y. Kitazawa and M. Ninomiya, Nucl. Phys. B **404**, 684 (1993).
19. M. Reuter, Phys. Rev. D57 (1998) 971.
20. W. Souma PTP **102**, 181 (1999).
21. A. Bedroya and C. Vafa, arXiv:1909.11063 [hep-th].
22. A. V. Frolov and L. Kofman, JCAP **0305**, 009 (2003).
23. N. Aghanim *et al.* [Planck Collaboration], arXiv:1807.06209 [astro-ph.CO].

Non-invertible topological duality defects
in 4-dimensional pure $\mathbb{Z}_2$ gauge theory

M. Koide*

Department of Physics, Graduate School of Science,
Osaka University, Toyonaka, Osaka 560-0043, Japan
* E-mail: mkoide@het.phys.sci.osaka-u.ac.jp

In recent years, the extension of the notion of symmetry using the picture of topological defects and its applications have been actively studied. One direction is the so-called non-invertible symmetry, in which non-invertible topological defects are treated as symmetry. We have constructed non-invertible topological defects from duality in 4D lattice pure $\mathbb{Z}_2$ gauge theory. This work is in collaboration with Y.Nagoya and S.Yamaguchi. [1]

Keywords: Non-invertible; symmetry; topological defect.

1. Introduction

Symmetry is one of the important tools in the non-perturbative analysis of quantum field theory. In recent years, the extension of the notion of symmetry and its applications have been actively studied [2]. In extended symmetry, topological operators are considered as symmetries. There are also several types of extended symmetries, depending on the properties of their topological operators. The one direction is called the non-invertible symmetry. The unitary topological operator corresponding to the ordinary symmetry has an inverse transformation due to the group structure of the symmetry. There are general topological operators that do not have such inversions, and these types of topological operators are treated as non-invertible symmetries.

The most famous concrete example of non-invertible symmetry is the duality defect associated with the Kramers-Wannie duality in the 2-dimensional Ising model. Since this duality defect vanishes when acting on a single spin operator, it is a non-invertible for which there is no inverse transformation. It is also known that there is a non-invertible symmetry defect called Verlinde line in the 2-dimensional RCFTs. Thus, while examples of non-invertible symmetry in 2-dimensions were known, examples of non-invertible symmetry in 4-dimensions were less understood than those in 2-dimensions.

One approach to constructing topological defects in lattices is the work done by Aasen, Mong, and Fendley(AMF).[3,4] In this paper, We explain how to use this AMF method to construct a high dimensional topological defect.[1]

2. 4-dimensional $\mathbb{Z}_2$ lattice gauge theory

In this paper, we examine the 4-dimensional pure $\mathbb{Z}_2$ lattice gauge theory. We prepare the 4-dimensional cubic lattice and assign a link variable $U_m = \pm 1$ to each link m. The partition function of this theory is given by,

$$Z = \sum_{\{U\}} \exp\left(K \sum_{i \in P} \prod_{m \in \square_i} U_m \right). \tag{1}$$

Here P is the set of all plaquettes, K is a positive constant parameter, and $\square_i$ is the set of four links included in the plaquette i. It was shown by Wegner that there is a duality in this theory, called the Kramers–Wannier–Wegner duality(KWW duality).[5] According to the KWW duality, these theories with the parameter K and $\hat{K}$ are equivalent when they satisfy the relation

$$\sinh 2K \sinh 2\hat{K} = 1. \tag{2}$$

The situation $K = \hat{K} = K_c$ is called self-dual point.

$$K_c = -\frac{1}{2} \log(-1 + \sqrt{2}). \tag{3}$$

Next, I explain how to extend the AMF method to work in the 4-dimensional pure $\mathbb{Z}_2$ lattice gauge theory. To consider the duality defect, we prepare two kinds of lattices as shown in Fig.1. To be more specific, we introduce coordinates (x_1, x_2, x_3, x_4) of $\mathbb{R}^4$. On this $\mathbb{R}^4$, we define two lattices $\Lambda := \{(x_1, x_2, x_3, x_4)|x_1, x_2, x_3, x_4 \in 2\mathbb{Z}\}$ and $\hat{\Lambda} := \{(x_1, x_2, x_3, x_4)|x_1, x_2, x_3, x_4 \in 2\mathbb{Z} + 1\}$ that are dual to each other. In each lattice, the line segment consisting of the two nearest points is called a link, and the smallest square consisting of the links is called a placket. A line segment connecting points on different lattices is not called a link.

We assign link variables $U_m = \pm 1$ to the links on Λ, but constant weights rather than variables to the links on the dual lattice $\hat{\Lambda}$ For this reason, lattice Λ given a variable is called an active lattice, and lattice $\hat{\Lambda}$

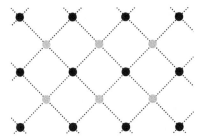

Fig. 1. Schematic figure of a lattice. In this figure, the lattice is depicted as 2-dimensional, but the actual lattice treated in this paper is a 4-dimensional one. The black lattice represents the lattice Λ, and the blue lattice represents the lattice $\hat{\Lambda}$. They are duals of each other.

given a constant weight is called an inactive lattice. This inactive lattice is an auxiliary lattice prepared to construct the duality defect.

Consider a 16-cell as the basic unit on these lattices as shown in Fig.2. A 16-cell consists of 16 tetrahedra such that each has one active link and one inactive link as edges. The surface of the 16-cell is homeomorphic to S^3.

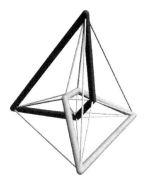

Fig. 2. 3-dimensional stereoscopic projection of a 16-cell. The black plaquette represent an active plaquette and the blue plaquette represent an inactive plaquette.

Since there is a one-to-one correspondence between active plaquettes and 16-cells, we assign Boltzmann weights to these 16-cells. If the four link variables in a 16-cell are $a_i = 0, 1$, $(i = 1, 2, 3, 4)$, we define the Boltzmann weight assigned to this 16-cell as

$$W(a_1, a_2, a_3, a_4) = \exp(K(-1)^{(a_1+a_2+a_3+a_4)}). \tag{4}$$

This definition is equivalent to the definition of Boltzmann weights in Eq. (1) ,if the link variables for the link m are identified as $U_m = (-1)^{a_m}$. We also assign a constant weight to each site or link, in addition to the link variable. We assign the weights for active links, active sites, inactive links, and inactive sites as s, l, $\bar{s}$, and $\bar{l}$, respectively. These weights are determined later to make the duality defect topological. The partition function is

$$
Z = \sum_{\{a\}} \left(\prod_{\substack{\text{active} \\ \text{sites}}} s \right) \left(\prod_{\substack{\text{active} \\ \text{links}}} l \right) \left(\prod_{\substack{\text{inactive} \\ \text{sites}}} \bar{s} \right) \left(\prod_{\substack{\text{inactive} \\ \text{links}}} \bar{l} \right)
$$
$$
\prod_{i \in C} W(a_{j_1(i)}, a_{j_2(i)}, a_{j_3(i)}, a_{j_4(i)}), \qquad (5)
$$

where C is the set of all 16-cells, and $j_1(i)$, $j_2(i)$, $j_3(i)$, and $j_4(i)$ are the four active links in the 16-cell i. a_j is the link variable assigned to the active link j.

3. Duality defect

We discuss topological KKW duality defects in the $\mathbb{Z}_2$ lattice gauge theory following AMF method[3,4]. This duality defects are 3-dimensional operators located at the boundary between two regions. The active lattice and the inactive lattice are swapped across the duality defect as shown in Fig.3. Because of this property, the building block of the duality defect is a tetrahedral prism with a doubled tetrahedron (see Fig.4). Each building block of duality defect contains two active links. Let $a, \tilde{a} = 0, 1$ be the link variables assigned to these two active links. We assign a weight $D(a, \tilde{a})$ $(a, \tilde{a} = 0, 1)$to this building block.

The entire duality defect is constructed by connecting these building blocks together. This defect is a 4-dimensional object in the lattice, but when the lattice spacing is set to zero, it becomes a 3-dimensional operator.

Since the duality transformation just changes the description of the theory and does not change the observables, we expect that such topological duality defects exist. In order to make the duality defect topological, we will impose a commutation relation on the defect and find its solution. From now on, we will consider the commutation relation on a single 16-cell. On the surface of a 16-cell, there are 16 tetrahedra. Consider an arrangement A of a certain duality defect. We assume that part of A occupies part of the surface of the 16-cell we are focusing on. We also consider a deformed

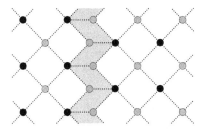

Fig. 3.   Schematic figure of the duality defect. A duality defect is placed at the boundary between the two regions. The active lattice (black dots) and the inactive lattice (blue dots) are swapped across the duality defect. The building block of the duality defect is a tetrahedral prism, depicted as a green parallelogram in this figure.

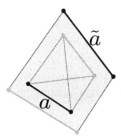

Fig. 4.   The building block of duality defects. the building block is a tetrahedral prism with a doubled tetrahedron. The 3-dimensional surface is composed of tetrahedrons each of which includes an active link and an inactive link.

configuration B in which the tetrahedrons which are not filled by A are filled and vice versa. Outside of a focused 16-cell, the configurations of A and B are assumed to be same. We assume that if the topologies of configuration A and B are the same, then the weights of those two configurations are equal. We call this relation commutation relation(Fig.5).

The commutation relation does not hold for some configurations, such as when the topology changes due to deformation. Therefore, we impose the following conditions on the commutation relation.

- The filled tetrahedrons on the 16-cell of A and B are both non-empty sets.
- The configurations A and B restricted on the surface of the focused 16-cell are both simply-connected.
- There is no connection such that the duality defects are connected only by sites or links.

In the following, we prepare some definitions to express the commutation

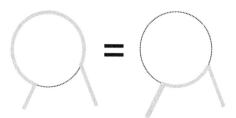

Fig. 5. Schematic figure of a commutation relation. The circle represents a 16-cell. A green line represents a duality defect. The commutation relation means that the value of the duality defect does not change when the topology is transformed without changing the topology.

relation as a mathematical expression. First, consider a 16-cell where the duality defect is not placed. Let $m = 1, 2, 3, 4$ be the four active links on the focused 16-cell, and $\tilde{n} = \tilde{1}, \tilde{2}, \tilde{3}, \tilde{4}$ be the four inactive links. Let the set of links be $M = \{1, 2, 3, 4\}$ and $\tilde{N} = \{\tilde{1}, \tilde{2}, \tilde{3}, \tilde{4}\}$, respectively. The active link and inactive link pair $(m, \tilde{n})$ corresponds one-to-one to the tetrahedron. The set of tetrahedra on the surface of a 16-cell U can be represented as follows

$$U = \{(m, \tilde{n}) | m = 1, 2, 3, 4, \ \tilde{n} = \tilde{1}, \tilde{2}, \tilde{3}, \tilde{4}\}. \tag{6}$$

Next, consider a 16-cell with a duality defect in configuration A. Suppose that $I \subset U$ is occupied by a duality defect. The duality defect is defined on a tetrahedral prism doubling the tetrahedron of the 16-cell surface, so an additional active link appears. Since this additional active link corresponds one-to-one to the original inactive link, we use $\tilde{n}$ as the label. And let $\tilde{E} = \{\tilde{n} | (m, \tilde{n}) \in I\}$ be the set of such additional active links. Thus, the weight of the duality defect on the focused 16-cell in coordination A can be expressed as

$$W(a_1, a_2, a_3, a_4) \prod_{(m, \tilde{n}) \in I} D(a_m, \tilde{a}_{\tilde{n}}). \tag{7}$$

In configuration B, the defect is placed at $\bar{I} = U \setminus I$. Therefore, the weight of the defect in configuration B becomes

$$W(\tilde{a}_{\tilde{1}}, \tilde{a}_{\tilde{2}}, \tilde{a}_{\tilde{3}}, \tilde{a}_{\tilde{4}}) \prod_{(m, \tilde{n}) \in \bar{I}} D(a_m, \tilde{a}_{\tilde{n}}). \tag{8}$$

Note that the Boltzmann weight variables are different for configurations A and B because the theories on the 16-cells are dual. We also consider $E = \{m | (m, \tilde{n}) \in \bar{I}\}$.

In this case, the commutation relation is written as follows

$$\sum_{M\backslash E} W(a_1,a_2,a_3,a_4)s^{\alpha_1}l^{\beta_1}\bar{s}^{\tilde{\alpha}_1}\bar{l}^{\tilde{\beta}_1} \prod_{(m,\tilde{n})\in I} D(a_m,\tilde{a}_{\tilde{n}})$$
$$= \sum_{\tilde{N}\backslash \tilde{E}} W(\tilde{a}_{\tilde{1}},\tilde{a}_{\tilde{2}},\tilde{a}_{\tilde{3}},\tilde{a}_{\tilde{4}})s^{\alpha_2}l^{\beta_2}\bar{s}^{\tilde{\alpha}_2}\bar{l}^{\tilde{\beta}_2} \prod_{(m,\tilde{n})\in \tilde{I}} D(a_m,\tilde{a}_{\tilde{n}}). \quad (9)$$

Here the number of active sites, active links, inactive sites, and inactive links in the configuration A are denoted by $\alpha_1,\beta_1,\tilde{\alpha}_1$, and $\tilde{\beta}_1$, respectively. The number of active sites, active links, inactive sites, and inactive links on the configuration B are denoted by $\alpha_2,\beta_2,\tilde{\alpha}_2$, and $\tilde{\beta}_2$, respectively. The sums $\sum_{M\backslash E},\sum_{\tilde{N}\backslash\tilde{E}}$ are defined as follows

$$\sum_{M\backslash E} := \prod_{m\in M\backslash E}\sum_{a_m=0,1}, \quad \sum_{\tilde{N}\backslash\tilde{E}} := \prod_{\tilde{n}\in\tilde{N}\backslash\tilde{E}}\sum_{\tilde{a}_{\tilde{n}}=0,1}. \quad (10)$$

In a physically sensible solution, these values satisfy

$$D(a,\tilde{a}) \neq 0, \quad l,s,\bar{l},\bar{s} > 0, \quad K\in\mathbb{R}, \quad K\neq 0. \quad (11)$$

By solving for the commutation relations, we obtained the following unique solution

$$D(a,\tilde{a}) = (-1)^{a\tilde{a}}, \quad (12)$$

$$l = \frac{1}{\sqrt{2}}, \quad s = \frac{1}{\sqrt{2}}, \quad (13)$$

$$\bar{l} = 1, \quad \bar{s} = 1, \quad (14)$$

$$K = K_c = -\frac{1}{2}\log(-1+\sqrt{2}), \quad (15)$$

$$W(a_1,a_2,a_3,a_4) = \exp(K(-1)^{(a_1+a_2+a_3+a_4)}). \quad (16)$$

Thus, a topological duality defect was obtained. Note that the value of K is determined to be the critical value K_c.

Finally, we show that this topological defect is non-invertible. To do this, consider the situation as $I = U$ as the placement of the duality defect in configuration A (see Fig.6). In this case, the topology of the duality defect changes with the deformation, so the commutation relation is not always satisfied. In fact, the calculation results in the following

$$\sum_{a_1,a_2,a_3,a_4=0,1} W(a_1,a_2,a_3,a_4)s^8l^8\bar{s}^8\bar{l}^8 \prod_{(m,\tilde{n})\in U} D(a_m,\tilde{a}_{\tilde{n}})$$
$$= \frac{1}{\sqrt{2}}W(\tilde{a}_{\tilde{1}},\tilde{a}_{\tilde{2}},\tilde{a}_{\tilde{3}},\tilde{a}_{\tilde{4}})s^4l^4\bar{s}^4\bar{l}^4. \quad (17)$$

For a symmetry defect placed on a closed manifold with no operator inserted inside, the weights are identical to the empty configuration[2]. However, for this duality defect, their ratio is $1/\sqrt{2}$. So, this defect is non-invertible.

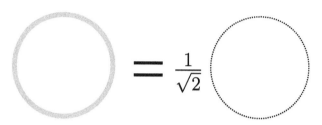

Fig. 6.   Schematic figure of Eq.(17). The circles represent the 16-cell. The green line represents the duality defect.

4. summary

We construct topological KWW duality defects in the $\mathbb{Z}_2$ lattice gauge theory following AMF method[3,4]. And this duality defect is non-invertible.

References

1. M. Koide, Y. Nagoya and S. Yamaguchi, Non-invertible topological defects in 4-dimensional $\mathbb{Z}_2$ pure lattice gauge theory (9 2021).
2. D. Gaiotto, A. Kapustin, N. Seiberg and B. Willett, Generalized Global Symmetries, *JHEP* **02**, p. 172 (2015).
3. D. Aasen, R. S. K. Mong and P. Fendley, Topological Defects on the Lattice I: The Ising model, *J. Phys. A* **49**, p. 354001 (2016).
4. D. Aasen, P. Fendley and R. S. K. Mong, Topological Defects on the Lattice: Dualities and Degeneracies (8 2020).
5. F. J. Wegner, Duality in Generalized Ising Models and Phase Transitions Without Local Order Parameters, *J. Math. Phys.* **12**, 2259 (1971).

Conifold transition and matter in non-split models in F-theory

R. Kuramochi

SOKENDAI (The Graduate University for Advanced Studies)
Tsukuba, Ibaraki, 305-0801, Japan
E-mail: rinto@post.kek.jp

We show that the split/non-split transition is, with some exceptions, a conifold transition at a D_{2k} ($k \geq 2$) or an E_7 codimension-two singularity, and these non-split models have local deformed conifolds. These clarify, in these cases, that the origin of non-local matter generation and that the deformations of conifold singularities correspond to diagram automorphisms of the expected simply-laced Dynkin diagrams in the split sides.

Keywords: F-theory, matter, singularity, resolution and deformation

1. Introduction

F-theory[1–3] is a geometrical framework of nonperturbative compactifications[a] of type IIB theory with general 7-branes. In six- or lower-dimensional F-theories, if a fiber type involves the condition that an exceptional curve splits into two irreducible ones, then the fiber type has two types, "split" or "non-split", depending on whether the exceptional curve can split globally or not[6]. In the split models, each intersection diagram of exceptional curves that arises after the resolution corresponds to the expected *ADE* Dynkin diagram implied by Kodaira's classification[7](Table 2). In the non-split models, since the two split exceptional curves are identified by monodromy, each *ADE* gauge symmetry in the split models is reduced to the corresponding non-simply-laced gauge symmetry by being subject to a projection by a diagram automorphism. The fiber type I_n ($n = 3, 4, \ldots$), I_n^* ($n = 0, 1, \ldots$), IV or IV^* can involve such identification of exceptional curves[6,8].

Non-split models have some puzzles which require a new understanding of the non-local generation of charged matter. In split models, if the matter fields locally exist at all codimension-two singularities[b], the number of the matter fields matches the anomaly-free condition[6]. This is one of the

[a] In F-theory, strong- and weak-coupling regions coexist on the base of the total space[4,5].
[b] In F-theory, there are 7-branes on codimension-one loci (in the base: codimension-two in the total space) where the elliptic fibers become singular. These codimension-one loci

reasons why the massless matter fields are localized at all intersections of 7-branes in F-theory. However, in non-split models, there is a puzzle that the anomaly-free spectrum and the naive counting are not consistent[6,12–16]. Moreover, at a D_{2k} $(k = 2, 3, \ldots)$ or an E_7 codimension-two singularity, some conifold singularities[c] remain after all codimension-one singularities are blown up in the split models, but not in the non-split models. In the split models, since we can yield new two-cycles by small resolutions of the conifold singularities[9,17], we can obtain an intersection diagram of exceptional curves which is different from one on a codimension-one singularity and explains the enhancement of the symmetry. In the M-theory dual, an M2-brane wrapping a new two-cycle generates local matter multiplets[18]. On the other hand, in the non-split models, no additional blow-up is required at these singularities, so the intersection diagrams remain the same and there is no new two-cycle where an M2-brane can wrap around[12].

This paper is based on our recent papers[12,19]: we discuss these puzzles[12] and show the split/non-split transition is a conifold transition at a D_{2k} or an E_7 codimension-two singularity with some exceptions (I^*_{2k-4})[19]. Since this implies these non-split models have local deformed conifolds[19], we can yield new three-cycles which are non-local in terms of an elliptic fibration. These clarify, in these cases, the "non-local" matter fields may satisfy the anomaly-free condition and the deformations of conifold singularities correspond to diagram automorphisms of the ADE Dynkin diagrams in the split sides.

2. Puzzles on matter generation in non-split models

In this section, we discuss non-split models using a non-split I_6 model as an example[12]. Specifically, we examine their matter spectra and the result of the resolution at a D_6 singularity of a non-split I_6 local equation. Thus, we show the non-split models have some puzzles regarding the generation of charged matter fields.

2.1. Puzzle on matter spectra near D_6 points in I_6 model

In this section, we consider the six-dimensional F-theory on an elliptically fibered Calabi-Yau three-fold over a Hirzebruch surface $\mathbb{F}_n$[1–3] in which the unbroken gauge symmetry is A_5 or C_3. These setups of F-theory compacti-

---

are called codimension-one singularities[6,8–11]. In general, there are intersections of the codimension-one loci in the base. These codimension-two loci, where codimension-one singularities are enhanced, correspond to intersections of 7-branes. So, The codimension-two loci are involved in matter generation and are called codimension-two singularities.
[c]This conifold singularity exists not only in the base but also in the elliptic fiber direction and appears naturally where matter generates without any special tuning of parameters.

fications are dual to the $E_8 \times E_8$ heterotic theories on $K3$ with $(12+n, 12-n)$ instantons[1-3]. In the heterotic dual, these massless spectra can be calculated using the index theorem (Table 1). They satisfy the anomaly-free condition for the E_8 with $12 + n$ instantons $n_H - n_V = 30n + 112$[6].

Table 1. The anomaly-free massless matter spectrum of heterotic theory on $K3$ in which the unbroken gauge symmetry is A_5 or C_3. Specifically, we consider the case where $12 + n$ instantons are distributed among $(8 + n - r, 4 + r)$ $(0 \leq r \leq n + 2)$ in (A_2, A_1) or (G_2, A_1).

Gauge group	Number of hypermultiplets: n_H
A_5	$\frac{r}{2}\mathbf{20} + (n + 2 - r)\mathbf{15} + (2n + 16 + r)\mathbf{6} + (3n - r + 21)\mathbf{1}$
C_3	$\frac{r}{2}\mathbf{14'} + (n + 1 - r)\mathbf{14} + (2n + 16 + \frac{3r}{2})\mathbf{6} + (4n - 2r + 23)\mathbf{1}$

In F-theory, an elliptic fibration can be described in Weierstrass form

$$y^2 = x^3 + fx + g, \tag{1}$$

where x and y are coordinates describing an elliptic curve. And f and g are holomorphic functions on a Hirzebruch surface (a $\mathbb{P}^1$ fibration over $\mathbb{P}^1$)

$$f(z, w) := \sum_{i=0}^{I} z^i f_{8+n(4-i)}(w), \qquad g(z, w) := \sum_{j=0}^{J} z^j g_{12+n(6-j)}(w),$$

where w and z are the affine coordinates along the base and $\mathbb{P}^1$ fiber, and I and J are the largest integers that satisfy $I \leq 8$, $8+n(4-I) \geq 0$ and $J \leq 12$, $12 + n(6 - J) \geq 0$, respectively. In this subsection, all subscripts denote the degree of the polynomial in w. We only consider $i \leq 4$, $j \leq 6$, and near $z = 0$ so that we focus on the E_8 with $12 + n$ instantons in the heterotic dual[1-3]. There are singular fibers over the zero loci of the discriminant

$$\Delta := 4f^3 + 27g^2 = 0. \tag{2}$$

These singularities can be classified by the vanishing orders z of f, g, and Δ on the codimension-one singularities (Kodaira's classification[7]). We obtain a split I_6 equation (as Table 1) from the equation (1)[6,12]:

$$
\begin{aligned}
f_{I_6^s}(z, w) := {}& -3t_r^4 h_{n-r+2}^4 + 6z t_r^3 h_{n-r+2}^2 H_{n-r+4} + 3z^2 t_r \left(2u_{r+4} h_{n-r+2}^2 \right. \\
& \left. - t_r H_{n-r+4}^2\right) + z^3 \left(t_r f_{n-r+8} - 3u_{r+4} H_{n-r+4}\right) + f_8 z^4, \tag{3}
\end{aligned}
$$

$$
\begin{aligned}
g_{I_6^s}(z, w) := {}& 2t_r^6 h_{n-r+2}^6 - 6z t_r^5 h_{n-r+2}^4 H_{n-r+4} - 6z^2 t_r^3 h_{n-r+2}^2 \left(u_{r+4} h_{n-r+2}^2 \right. \\
& \left. - t_r H_{n-r+4}^2\right) + z^3 t_r^2 \left(9u_{r+4} h_{n-r+2}^2 H_{n-r+4} - t_r f_{n-r+8} h_{n-r+2}^2 \right. \\
& \left. - 2t_r H_{n-r+4}^3\right) + z^4 \left(3u_{r+4}^2 h_{n-r+2}^2 + t_r^2 f_{n-r+8} H_{n-r+4} \right. \\
& \left. - f_8 t_r^2 h_{n-r+2}^2 - 3t_r u_{r+4} H_{n-r+4}^2\right) \\
& + z^5 \left(f_8 t_r H_{n-r+4} + u_{r+4} f_{n-r+8}\right) + g_{12} z^6, \tag{4}
\end{aligned}
$$

Table 2. Kodaira's classification of singularities of an elliptic surface[7].

ord(f)	ord(g)	ord(Δ)	Fiber type	Singularity type G
≥ 0	≥ 0	0	smooth	none
0	0	n	I_n	A_{n-1}
≥ 1	1	2	II	none
1	≥ 2	3	III	A_1
≥ 2	2	4	IV	A_2
2	≥ 3	$n+6$	I_n^*	D_{n+4}
≥ 2	3	$n+6$	I_n^*	D_{n+4}
≥ 3	4	8	IV^*	E_6
3	≥ 5	9	III^*	E_7
≥ 4	5	10	II^*	E_8
≥ 4	≥ 6	≥ 12	non-minimal	–

Table 3. Massless matter content in the I_6^s model of F-theory, if the matter fields locally exist at all codimension-two singularities. In I_6^{ns}, each representation of I_6^s simply decomposes into an irreducible representation of C_3 at each zero locus.

Enhancement in I_6^s	I_6^s (A_5)		I_6^{ns} (C_3)	
	Matter rep.	Multiplicity	Matter rep.	Multiplicity
IV^* (E_6)	$\frac{1}{2}\mathbf{20}$	r	$\frac{1}{2}(\mathbf{14'}+\mathbf{6})$	r
I_2^* (D_6)	$\mathbf{15}$	$n+2-r$	$\mathbf{14}$	$2n+4-2r$
I_7 (A_6)	$\mathbf{6}$	$2n+16+r$	$\mathbf{6}$	$2n+16+r$
	$\mathbf{1}$	$3n+21-r$	$\mathbf{1}$	$4n+23-2r$

where t_r, h_{n-r+2}, H_{n-r+4}, u_{r+4}, f_{n-r+8}, g_8, and g_{12} are the sections of appropriate line bundles over the base $\mathbb{P}^1$. And the discriminant (2) is

$$\Delta_{I_6^s} = z^6 t_r^3 h_{n-r+2}^4 P_{2n+r+16} + z^7 t_r^2 h_{n-r+2}^2 Q_{3n+20} + z^8 R_{4n+24} + O(z^9), \quad (5)$$

where $P_{2n+r+16}$, Q_{3n+20}, and R_{4n+24} are some non-factorizable polynomials. At the zero loci of t_r, h_{n-r+2}, and $P_{2n+r+16}$, the codimension-one singularity is enhanced to IV^* (E_6), I_2^* (D_6), and I_7 (A_6), respectively. These zero loci are the codimension-two singularities. In general, any two of t_r, h_{n-r+2}, and $P_{2n+r+16}$ will not be zero at the common locus, so we will get the matter content (Table 3) from (3), (4), and (5). It can be verified from Tables 1 and 3 that if the matter locally exists at all codimension-two singularities, then the anomaly-free condition is satisfied.

A non-split I_6 equation, which reduces the gauge group from A_5 to C_3, is obtained by the replacement of the section $h_{n-r+2}^2 \to h_{2n-2r+4}$ (not changing t_r and $P_{2n+r+16}$) in the split I_6 equation since the two split exceptional curves can no longer be factorized[6]. Thus, we consider each matter representation of A_5 simply decomposes into an irreducible representation of C_3, respectively (Table 3). In this case, the number of the zero loci where the codimension-one singularity is enhanced to I_2^* is doubled, but we cannot assign it to two separate zero loci since **14** of C_3 is not a pseudo-real but a real representation. Therefore, it is difficult to realize the local generation of matter at all codimension-two singularities in the non-split models.

2.2. *Intersections of exceptional curves in non-split I_6*

As we will see explicitly in the next section, we examine, in the non-split I_6 model with a D_6 codimension-two singularity, the intersection diagrams of exceptional curves near $h_{2n-2r+4} = w = 0$ (in Section 3, $b_{2,0} = w = 0$).

At a point of $w \neq 0$, in the split I_6 model, we have five exceptional curves $\mathcal{C}_{p_1}^{\pm}$, $\mathcal{C}_{p_2}^{\pm}$, and $\mathcal{C}_{p_3}$ (the top of Figure 1). In the non-split I_6 model, since the two split exceptional curves $\mathcal{C}_{p_1}^{\pm}$ cannot be factorized in terms of the polynomial ring of w and replace each other at $w = 0$, $\mathcal{C}_{p_1}^{\pm}$ are identified; the same applies to $\mathcal{C}_{p_2}^{\pm}$. This corresponds to the diagram automorphism of the Dynkin diagram: $\mathcal{C}_{p_i} := \frac{1}{2}(\mathcal{C}_{p_i}^+ + \mathcal{C}_{p_i}^-)$ $(i = 1, 2)$ (the middle of Figure 1).

At $w = 0$, no conifold singularity remains after the codimension-one singularity is blown up[12]. Thus, as in the previous works[9–11], by lifting up the exceptional curves $\mathcal{C}$'s and δ's (as $(x, y, z) \to (x_1 z, y_1 z, z)$ in Section 3.2), we obtain $\mathcal{C}_{p_1} \to \delta_{p_1}$, $\mathcal{C}_{p_2} \to \delta_{p_2}$, $\mathcal{C}_{p_3} \to \delta_{p_3}$ (the bottom of Figure 1).

Therefore, in the non-split models, even at the codimension-two singularities, the intersection diagrams of exceptional curves remain the same. Thus there is no new two-cycle where an M2-brane can wrap around.

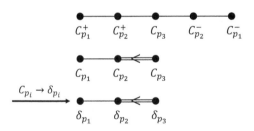

Fig. 1.   Intersection diagrams of the exceptional curves (based on Ref. 12): (Top) $w \neq 0$ before the identification; (Middle) $w \neq 0$ after the identification; (Bottom) $w = 0$.

3. Split/non-split transition as a conifold transition

In this section, to show the split/non-split transition is a conifold transition in the I_{2k} model, we will perform a concrete blowing-up process of a codimension-one singularity, near a D_{2k} codimension-two singularity. After that, we will show the non-split models have nontrivial three-cycles; we will discuss the origin of non-local matter in the non-split models[19]. Finally, we will discuss all other fiber types in which the non-split models exist[19].

3.1. *The local equation*

A local elliptic fibration with a codimension-one singularity along $(x, y, z) = (0, 0, 0)$ can be described in "Degline form"[20]:

$$\Phi(x, y, z, w) = -y^2 + x^3 + \frac{b_2}{4}x^2 + \frac{b_4}{2}x + \frac{b_6}{4} = 0,$$

$$b_8 := \frac{1}{4}(b_2 b_6 - b_4^2), \tag{6}$$

$$\Delta = \frac{1}{16}\left(b_2^2 b_8 - 9b_2 b_4 b_6 + 8b_4^3 + 27b_6^2\right),$$

where b_j is a section. b_j is expanded as $b_j = b_{j,0} + b_{j,1}z + \cdots + b_{j,j}z^j$ ($j = 2, 4, 6$), where $b_{j,k}$ is a $((j-k)n + 2j)$-th degree polynomial in w.

There is a relationship between "Degline form" (6) and Weierstrass form (1) in the previous section as follows: $f = -\frac{1}{48}(b_2^2 - 24b_4)$, $g = \frac{1}{864}\left(b_2^3 - 36b_2 b_4 + 216b_6\right)$. f and g are sections of the same line bundle as b_4 and b_6, respectively. And then, in the dP$_9$ fibration they are expanded as $f = f_{4,0} + f_{4,1}z + \cdots + f_{4,4}z^4$, $g = g_{6,0} + g_{6,0}z + \cdots + g_{6,6}z^6$, where $f_{4,k}$, $g_{6,k}$ are written as $f_{8+n(4-k)}$, $g_{12+n(6-k)}$ in the previous section.

The conditions for singularities of the split and non-split fiber types are summarized in Table 4[19]. Thus, substituting $b_2 := 4[(w-\epsilon)(w+\epsilon) - z]$, $b_4 := 2z^k$, and $b_6 := 4z^{2k}$ into equation (6), we obtain the local equation with an I_{2k} codimension-one singularity along $(x, y, z) = (0, 0, 0)$ for arbitrary w:

$$\Phi(x, y, z, w) = -y^2 + x^3 + [(w - \epsilon)(w + \epsilon) - z]x^2 + z^k x + z^{2k} = 0,$$

$$b_8 = z^{2k}\left[4(w - \epsilon)(w + \epsilon) - 1 - 4z\right],$$

$$\Delta = z^{2k}\left[(w - \epsilon)^2(w + \epsilon)^2(4w^2 - 4\epsilon^2 - 1)\right. \tag{7}$$

$$- 2(w - \epsilon)(w + \epsilon)(6w^2 - 6\epsilon^2 - 1)z$$

$$\left. + (12w^2 - 12\epsilon^2 - 1)z^2 + O(z^3)\right],$$

where $\epsilon \in \mathbb{R}$ is the parameter for a split/non-split transition since $b_{2,0}$ can be factorized when $\epsilon = 0$ and not when $\epsilon \neq 0$. In this section, for simplicity, we consider $k \geq 3$, but we can make the same discussion for $k = 2$.

3.2. *Resolutions of I_{2k} codimension-one singularities*

We consider the resolution of a codimension-one singularity[9–11] of the local equation (7) along $(x, y, z) = (0, 0, 0)$ for arbitrary w. This equation (7) also has one D_{2k} codimension-two singularity at $w = +\epsilon$ and another one at $w = -\epsilon$ in the non-split side, while it has one at $w = 0$ in the split side. So, we obtain the exceptional curves $\mathcal{C}$'s at $w \neq \pm\epsilon$ and δ's at $w - \pm\epsilon$.

We replace $(x, y, z) = (0, 0, 0)$ for some fixed w with a $\mathbb{P}^2$, by replacing

Table 4. Singularities of the split and non-split fiber types. I_{2k+1}^{os} denotes the "over-split" type which is explained in Ref. 19.

Kodaira's fiber type	ord(b_2)	ord(b_4)	ord(b_6)	ord(b_8)	ord(Δ)	Additional constraint(s)	Split/non-split fiber type
$I_{2k}(k \geq 2)$	0	k	$2k$	$2k$	$2k$	$b_{2,0} = c_{1,0}^2$	I_{2k}^s
						$b_{2,0}$ generic	I_{2k}^{ns}
$I_{2k+1}(k \geq 1)$	0	k	$2k$	$2k+1$	$2k+1$	$\begin{cases} b_{2,0} = c_{1,0}^2 \\ b_{4,k} = c_{1,0}c_{3,k} \\ b_{6,2k} = c_{3,k}^2 \end{cases}$	I_{2k+1}^s
						$\begin{cases} b_{2,0} \text{ generic} \\ b_{4,k} = b_{2,0}c_{2,k} \\ b_{6,2k} = b_{2,0}c_{2,k}^2 \end{cases}$	I_{2k+1}^{ns}
						$\begin{cases} b_{2,0} = c_{1,0}^2 \\ b_{4,k} = c_{1,0}^2 c_{2,k} \\ b_{6,2k} = c_{1,0}^2 c_{2,k}^2 \end{cases}$	I_{2k+1}^{os}
I_0^*	1	2	3	4	6	$\begin{cases} b_{2,1} = 4(p_{2,1} + q_{2,1} + r_{2,1}) \\ b_{4,2} = 2(p_{2,1}q_{2,1} + q_{2,1}r_{2,1} \\ \qquad\quad + r_{2,1}p_{2,1}) \\ b_{6,3} = 4p_{2,1}q_{2,1}r_{2,1} \end{cases}$	I_0^{*s}
						$\begin{cases} b_{2,1} = 4(p_{2,1} + q_{2,1}) \\ b_{4,2} = 2(p_{2,1}q_{2,1} + r_{4,2}) \\ b_{6,3} = 4p_{2,1}r_{4,2} \end{cases}$	I_0^{*ss}
						$b_{2,1}, b_{4,2}, b_{6,3}$ generic	I_0^{*ns}
$I_{2k-3}^*(k \geq 2)$	1	$k+1$	$2k$	$2k+1$	$2k+3$	$b_{6,2k} = c_{3,k}^2$	I_{2k-3}^{*s}
						$b_{6,2k}$ generic	I_{2k-3}^{*ns}
$I_{2k-2}^*(k \geq 2)$	1	$k+1$	$2k+1$	$2k+2$	$2k+4$	$b_{8,2k+2} = c_{4,k+1}^2$	I_{2k-2}^{*s}
						$b_{8,2k+2}$ generic	I_{2k-2}^{*ns}
IV	1	2	2	3	4	$b_{6,2} = c_{3,1}^2$	IV^s
						$b_{6,2}$ generic	IV^{ns}
IV^*	2	3	4	6	8	$b_{6,4} = c_{3,2}^2$	IV^{*s}
						$b_{6,4}$ generic	IV^{*ns}

$\mathbb{C}^3$ with $\hat{\mathbb{C}}^3 = \{((x,y,z),(\xi : \eta : \zeta)) \in \mathbb{C}^3 \times \mathbb{P}^2 | (x : y : z) = (\xi : \eta : \zeta)\}$. We will be blowing up the codimension-one singularity in inhomogeneous coordinates defined in the three different affine patches of $\mathbb{P}^2$, for example, $(x : y : z) = (\xi : \eta : \zeta) = (x_1 : y_1 : 1)$ ($1_z, z \neq 0$). Then, to replace $\mathbb{C}^3$ with $\hat{\mathbb{C}}^3$, we simply replace (x, y, z) with $(x_1 z, y_1 z, z)$ in the equation (7). To not change the canonical class, the equation after the blow-up is defined as follows: $z^{-2}\Phi(x_1 z, y_1 z, z, w) =: \Phi_z(x_1, y_1, z, w) = 0$. The other patches are also similar. In this (7) case, the codimension-one singularity remains along $(x_j, y_j, z) = (0, 0, 0)$ ($j = 1, \cdots, k-1$). Thus, we perform similar processes k times for the resolution of the codimension-one singularity.

Performing these operations, we obtain the j times ($j = 1, \cdots, k$) blown-up equation in patch $j_{\underbrace{z \cdots z}_{j}}$ is

$$\Phi_{\underbrace{z\cdots z}_{j}}(x_j, y_j, z, w) = -y_j^2 + z^j x_j^3 + [(w-\epsilon)(w+\epsilon) - z]\,x_j^2 + z^{k-j}x_j + z^{2(k-j)}$$

$$= 0. \tag{8}$$

These equations have codimension-one singularities along $(x_j, y_j, z) = (0,0,0)$ in $j = 1, \cdots, k-1$, and not in $j = k$. In the non-split side ($\epsilon \neq 0$), we also obtain the exceptional curves at $w \neq \pm\epsilon$ and $w = \pm\epsilon$ as follows:

$$\begin{aligned} C_{p_j} &: z = 0, \quad y_j^2 = (w-\epsilon)(w+\epsilon)x_j^2 \\ \delta_{p_j} &: z = 0, \quad y_j = 0, \quad w = \pm\epsilon \end{aligned} \quad (j = 1, \cdots, k-1), \tag{9}$$

$$\begin{aligned} C_{p_k} &: z = 0, \quad y_k^2 = (w-\epsilon)(w+\epsilon)x_k^2 + x_k + 1 \\ \delta_{p_k} &: z = 0, \quad y_k^2 = x_k + 1, \quad w = \pm\epsilon \end{aligned} \quad (j = k). \tag{10}$$

3.3. Local deformed conifolds in non-split models and split/non-split transition as a conifold transition

To see conifold singularities, we consider the j times ($j = 1, \cdots, k$) blown-up equation in the patch of $j_{\underbrace{z\cdots z}_{j-1}x}$

$$0 = \Phi_{\underbrace{z\cdots z}_{j-1}x}(x_{j-1}, y_j, z_j, w) = -y_j^2 - z_j x_{j-1} + w^2 - \epsilon^2 + z_j^{j-1}x_{j-1}^j$$

$$+ z_j^{k-j+1}x_{j-1}^{(k-j)} + z_j^{2(k-j+1)}x_{j-1}^{2(k-j)}, \tag{11}$$

where (x_0, y_0, z_0) means (x, y, z). In the split side ($\epsilon = 0$), we can confirm that there are k conifold singularities at a D_{2k} codimension-two singularity:

$$\begin{aligned} v_{q_1} &: \; (x, y_1, z_1, w) = (0, 0, 1, 0), \\ v_{q_j} &: \; (x_{j-1}, y_j, z_j, w) = (0, 0, 0, 0) \quad (j = 2, \cdots, k-1), \\ v_{r_{k-1}} &: \; (x_{k-2}, y_{k-1}, z_{k-1}, w) = (0, 0, 1, 0). \end{aligned} \tag{12}$$

Thus, the intersection diagrams of exceptional curves are obtained by blowing up all conifold singularities (small resolutions), which explain the enhancements of the symmetries. The smooth split I_{2k} model corresponds to the resolved side of a conifold transition.

In the non-split side ($\epsilon \neq 0$), by appropriate variable transformations of equations (11) and considering only their lowest order, we obtain

$$y_j^2 + z_j x_{j-1} + w^2 - \epsilon^2 = 0 \quad (j = 1, \cdots, k-1). \tag{13}$$

These are local deformed conifolds[19]. Thus, it is natural that there is no conifold singularity in the non-split models; there are three-cycles instead of two-cycles. The smooth non-split I_{2k} model corresponds to the deformed side of a conifold transition.

The split/non-split transition parameter ϵ also plays the role of a conifold transition parameter, so the split/non-split transition is a conifold transition at a D_{2k} codimension-two singularity in the I_{2k} model. This clarifies the deformation of conifold singularity is associated with a diagram automorphism of the expected simply-laced Dynkin diagram in the split side.

Moreover, in the non-split models, it turns out that new S^3's, which are non-local in terms of the elliptic fibration, are generated. Thus, this can be regarded as the origin of non-local matter which cannot be generated from two-cycles. Since it is important to consider two D_{2k} codimension-two singularities as a pair to find these three-cycles, the matter generated from three-cycles may satisfy the anomaly-free condition. However, this requires a new way to generate matter fields from three-cycles with finite size. We will leave this issue that should be clarified in the future.

3.4. All other fiber types with non-split models

Similar results are obtained in IV and I_{2k-5}^* ($k \geq 3$) at a D_{2k} point, and in IV^* at an E_7 point [19]. In I_{2k-1} ($k \geq 2$), a split model has no D_{2k} point in general. But we get a D_{2k} point if we adjust a D_{2k-1} and an A_{2k-1} to overlap at the same point. We call such a specially tuned fiber type an "over-split" [19]. At this D_{2k} point, we obtain a similar discussion through an "over-split" model. In I_{2k-2}^* ($k \geq 1$), conifold singularities do not appear after the resolution of the codimension-one singularity. Therefore, this is the only case where similar arguments cannot be applied [19].

4. Conclusion

In this paper, we have considered the non-split models have two puzzles on matter generation, using I_6 specifically: first, if the matter fields locally exist and matter representations in the split models simply decompose into irreducible representations of the corresponding non-simply-laced gauge groups at all codimension-two singularities, the number of matter fields mismatches the anomaly-free condition [6,12–16]; second, there is no conifold singularity at all D_{2k} or E_7 points after all codimension-one singularities are blown up, so the intersection diagrams of exceptional curves remain the same [12]. Next, using I_{2k} specifically, we have shown the split/non-split transition is a conifold transition at a D_{2k} codimension-two singularity, and nontrivial three-cycles S^3 are generated in the non-split models. These have been clarified, in these cases, the deformations of conifold singularities are associated with diagram automorphisms of the corresponding simply-laced Dynkin diagrams in the split sides. Moreover, we have discussed how these solve the above puzzles [19]. However, in the I^*_{2k-2} models, we cannot discuss the above. Also, we need a new understanding of matter generation due to three-cycles. These are issues that need to be clarified in the future. The analyses based on the deformations [21,22] could help in solving these issues.

Acknowledgements

We thank H. Itoyama, Y. Imamura, K. Sakai, S. Mizoguchi, T. Tani, Y. Kimura, and H. Otsuka for useful discussions.

References

1. C. Vafa, Nucl. Phys. B **469** (1996), 403.
2. D. R. Morrison and C. Vafa, Nucl. Phys. B **473** (1996), 74.
3. D. R. Morrison and C. Vafa, Nucl. Phys. B **476** (1996), 437.
4. S. Fukuchi, N. Kan, S. Mizoguchi and H. Tashiro, Phys. Rev. D **100** (2019) no. 12, 126025.
5. S. Fukuchi, N. Kan, R. Kuramochi, S. Mizoguchi and H. Tashiro, Phys. Lett. B **803** (2020), 135333.
6. M. Bershadsky, K. Intriligator, S. Kachru, D.R. Morrison, V. Sadov and C. Vafa, Nucl. Phys. B481 (1996), 215-252.
7. K. Kodaira, Ann. of Math. **77** (1963), 563.
8. S. Katz, D. R. Morrison, S. Schafer-Nameki and J. Sully, JHEP **08** (2011), 094.
9. D. R. Morrison and W. Taylor, JHEP **01** (2012), 022.
10. N. Kan, S. Mizoguchi and T. Tani, JHEP **08**(2020), 063.
11. C. Lawrie and S. Schäfer-Nameki, JHEP **04** (2013), 061.
12. R. Kuramochi, S. Mizoguchi and T. Tani, Progress of Theoretical and Experimental Physics, 2022;, ptac022, doi:10.1093/ptep/ptac022.
13. P. Arras, A. Grassi and T. Weigand, J. Geom. Phys. **123** (2018), 71-97.
14. M. Esole, P. Jefferson and M. J. Kang, arXiv:1704.08251 [hep-th].
15. M. Esole and M. J. Kang, JHEP **02** (2019), 091.
16. M. Esole and P. Jefferson, arXiv:1910.09536 [hep-th].
17. M. Esole and S. T. Yau, Adv. Theor. Math. Phys. **17** (2013) no.6, 1195.
18. S. H. Katz and C. Vafa, Nucl. Phys. B **497** (1997), 146.
19. R. Kuramochi, S. Mizoguchi and T. Tani, arXiv:2108.10136 [hep-th].
20. P. Deligne, Lecture Notes in Math., Vol. 476, Springer, Berlin, 1975.
21. A. Grassi, J. Halverson and J. L. Shaneson, JHEP **1310** (2013), 205.
22. A. Grassi, J. Halverson and J. L. Shaneson, Commun. Math. Phys. **336** (2015) no.3, 1231-1257.

Entanglement entropy in Schwarzschild spacetime

Yoshinori Matsuo

Department of Physics, Kyoto University,
Kitashirakawa, Kyoto 606-8502, Japan

In this work, we discuss the entanglement entropy in the Schwarzschild space-time, and its relation to the vacuum state of matter fields. Recently, it was proposed that there is either large violation of the additivity conjecture in quantum information theory or disentangled states of black holes in the AdS/CFT correspondence. Here, we consider the additivity conjecture in the Schwarzschild spacetime. Usually, the entanglement entropy is calculated assuming that the vacuum state is in the Hartle-Hawking vacuum. We discuss the entanglement entropy in the Schwarzschild spacetime, and its relation to the vacuum state of matter fields. We show that the other vacua than the Hartle-Hawking vacuum should be taken into consideration in order to consider the additivity conjecture.

1. Introduction and summary

Recently, it was proposed that there is either large violation of the additivity conjecture in quantum information theory or a set of disentangled states in the black hole spacetime [1]. A simplest statement of the additivity conjecture[2–8] says that the minimum output entropy

$$S_{\min}(\mathcal{N}) = \min_{\rho \in \mathcal{A}} S(\mathcal{N}(\rho)) \tag{1}$$

of two quantum channels $\mathcal{N}_1$ and $\mathcal{N}_2$ which map states in Hilbert spaces $\mathcal{A}_1$ or $\mathcal{A}_2$ to those in another Hilbert spaces $\mathcal{B}_1$ or $\mathcal{B}_2$ satisfies the additivity condition;

$$S_{\min}(\mathcal{N}_1 \otimes \mathcal{N}_2) = S_{\min}(\mathcal{N}_1) + S_{\min}(\mathcal{N}_2) , \tag{2}$$

where $S(\rho)$ is the con Neumann entropy of the density matrix ρ.

Hayden and Penington studied the additivity conjecture in the AdS/CFT correspondence [1]. Two Hilbert spaces $\mathcal{A}_1$ and $\mathcal{A}_2$ are identified to those of two conformal field theories which correspond to two boundaries of the AdS black hole spacetime. They considered quantum channels $\mathcal{N}_1$ and $\mathcal{N}_2$ which take the partial trace in each CFT. The output entropy of the channels are the entanglement entropy, which can be calculated by using the Ryu-Takayanagi formula [9, 10]. Then, they argued that the entanglement entropy does not satisfy the additivity condition (2). Hayden and Penington proposed that there is either large violation of the additivity conjecture or a set of disentangled quantum states which correspond to disconnected geometries in the gravity side [1].

Here, we consider the additivity conjecture in the Schwarzschild spacetime. We consider the quantum states of matters in two exteriors of the horizon in the Schwarzschild spacetime (See Fig. 1(left)). In a similar fashion to the case of the AdS/CFT correspondence, we take the partial trace of the states to obtain the entanglement entropy of the Hawking radiation. The entanglement entropy is usually calculated by assuming that the state is given by the Hartle-Hawking vacuum. The total entanglement entropy of the Hawking radiation is smaller than the sum of the entanglement entropy in each exterior, and hence, the additivity conjecture is not satisfied.

This is because we considered only the Hartle-Hawking vacuum, and the other vacua should be taken into account. In the other static vacua than the Hartle-Hawking vacuum, the quantum energy-momentum tensor becomes very large near the horizon. By solving the semi-classical Einstein equation, we find that two exteriors of the horizon are disconnected (Fig. 1(right)), as is proposed by Hayden and Penington. Thus, it is important to consider general static vacua in order to reproduce the additivity conjecture in the Schwarzschild spacetime.

Fig. 1. The Penrose diagram of the Schwarzschild spacetime in the Hartle-Hawking vacuum (left) and the semi-classical Schwarzschild spacetime in the other static vacua (right).

This paper is organized as follows. In Sec. 2, we briefly review the additivity conjecture in the AdS/CFT correspondence. In Sec. 3, we calculate the entanglement entropy in the Schwarzschild spacetime. In Sec. 4, we solve the semi-classical Einstein equation to obtain the semi-classical Schwarzschild solution. In Sec. 5, we consider the additivity conjecture in the Schwarzschild spacetime. This contribution is based on [11].

2. Additivity conjecture in AdS/CFT correspondence

The AdS/CFT correspondence relates the AdS spacetime with conformal field theory. In the cases of the eternal black holes in asymptotically AdS spacetime, the geometry have two boundaries, which correspond to two conformal field theories. The Einstein-Rosen bridge which connects two exteriors of the event horizon implies that states in two conformal field theories are entangled with each other. This entangled state is interpreted as the thermofield double state at the Hawking temperature,

$$|\psi\rangle = \sum e^{-\beta E_n/2} |n_+\rangle |n_-\rangle \ . \tag{3}$$

We consider the additivity conjecture for states in two CFTs $\mathcal{A}_1$ and $\mathcal{A}_2$. Two quantum channels $\mathcal{N}_i$ take partial trace in each CFT and maps states in total system of each CFT to those in subregions B_1 and B_2. The output entropy of the channels are nothing but the entanglement entropy of region $\mathcal{B}_i$,

$$S(\mathcal{N}_i) = S(B_i) \ , \qquad S(\mathcal{N}_1 \otimes \mathcal{N}_2) = S(B_1 \cup B_2) \ . \tag{4}$$

By using the Ryu-Takayanagi formula, the entanglement entropy of the region B in CFT is given by the area of the minimal area surface γ_B whose boundaries are anchored at the boundaries of B in the AdS boundary,

$$S(B) = \frac{\text{Area}(\gamma_B)}{4G_N} \ . \tag{5}$$

Fig. 2.   The geometry of a time slice in the AdS black hole spacetime and the Ryu-Takayanagi surface for B_1 or B_2 (left), and that for $B_1 \cup B_2$ (right).

For the entanglement entropy of B_1, the minimal surface lies in one exterior of the event horizon (Fig. 2 (left)). For the entanglement entropy of $B_1 \cup B_2$, the minimal surface extends between two boundaries through the Einstein-Rosen bridge (Fig. 2 (right)) [12]. Thus, we have

$$S(B_1 \cup B_2) < S(B_1) + S(B_2) \ . \tag{6}$$

Therefore, the additivity conjecture is violated by the typical states of black holes in the asymptotically AdS spacetime.

3. Entanglement entropy in Hartle-Hawking vacuum

Now, we consider the additivity conjecture for states of matters in the Schwarzschild spacetime. The Schwarzschild metric is given by

$$ds^2 = -\left(1 - \frac{r_h}{r}\right) dt^2 + \left(1 - \frac{r_h}{r}\right)^{-1} dr^2 + r^2 d\Omega^2 . \tag{7}$$

The quantum channels $\mathcal{N}_i$ map states in total system into those of the Hawking radiation, which are defined as states in the region R, or equivalently in $r > b$. The entanglement entropy of the Hawking radiation is given by [13–17]

$$S = \sum_{\partial R,\ \partial I} \frac{\text{Area}}{4G_N} + \sum_{\partial R,\ \partial I} S_{\text{matter}}^{(\text{non-local})} , \tag{8}$$

where

$$S_{\text{matter}}^{(\text{non-local})} = (-1)^{i-j+1} \frac{c}{6} \sum_{i \neq j} \log |(U_i - U_j)(V_i - V_j)| + \frac{c}{12} \sum_i \log |g_{UV}(x_i)| , \tag{9}$$

and the summation is over the endpoints of region R and the island I [18–23] (if there is the island), which are labeled so that $i - j = \pm 1$ for neighboring labels. For the Hartle-Hawking vacuum, the coordinates in (9) should be the Kruskal coordinates.

We first consider the entanglement entropy of the Hawking radiation in two exteriors, $R_1 \cup R_2$. Before the Page time [24, 25], the configuration without the island dominates, and the entanglement entropy is calculated as [26]

$$S = \frac{2\pi b^2}{G_N} + \frac{c}{6} \log \left[\frac{16 r_h^2 (b - r_h)}{b} \cosh^2 \frac{t_b}{2r_h} \right] . \tag{10}$$

After the Page time, the configuration with an island dominates. The position of the island is determined so that the entanglement entropy is extremized. The entanglement entropy is evaluated as [26]

$$S \simeq \frac{2\pi r_h^2}{G_N} + \frac{2\pi b^2}{G_N} + \frac{c}{6} \left[\log \left(\frac{16 r_h^3 (b - r_h)^2}{b} \right) + \frac{b - r_h}{r_h} \right] . \tag{11}$$

Next, we calculate the entanglement entropy of the Hawking radiation in one of two exteriors, R_1. There is no configuration with islands. The entanglement entropy is calculated by introducing the IR cut-off Λ and is given by [27]

$$S = \frac{\pi b^2}{G_N} + \frac{c}{6} \log \Lambda , \tag{12}$$

which implies the entanglement entropy is infinitely large after taking $\Lambda \to \infty$. Therefore, the entanglement entropy of the Hawking radiation in the Hartle-Hawking vacuum does not satisfy the additivity condition,

$$S(R_1 \cup R_2) \ll S(R_1) + S(R_2) . \tag{13}$$

This apparent violation of the additivity conjecture is because we consider only the Hartle-Hawking vacuum. The Hartle-Hawking vacuum is given in the form of the thermofield double state in terms of the states in each of two exteriors. By using these states in each exterior, we can construct other static vacuum states. Since we need to consider the state with the minimum output entropy, the other vacua than the Hartle-Hawking vacuum must be taken into account.

4. Semi-classical Schwarzschild spacetime

In the other vacuum states than the Hartle-Hawking vacuum, the energy-momentum tensor on the classical Schwarzschild spacetime diverges at the event horizon. This implies that quantum effects becomes very important near the horizon, and we should solve the semi-classical Einstein equation,

$$R_{\mu\nu} - \frac{1}{2} g_{\mu\nu} R = 8\pi G_N \langle T_{\mu\nu} \rangle . \tag{14}$$

In the s-wave approximation, the expectation value of the quantum energy-momentum tensor is completely fixed by the conservation law and Weyl anomaly. For the most general spherically symmetric spacetime,

$$ds^2 = -C(u,v) du\, dv + r^2(u,v) d\Omega^2 , \tag{15}$$

the energy-momentum tensor is calculated as [28, 29]

$$\langle T_{uu} \rangle = -\frac{c}{48\pi^2 r^2} C^{1/2} \partial_u^2 C^{-1/2} + \frac{c\,\kappa^2}{192\pi^2 r^2} , \tag{16}$$

$$\langle T_{vv} \rangle = -\frac{c}{48\pi^2 r^2} C^{1/2} \partial_v^2 C^{-1/2} + \frac{c\,\kappa^2}{192\pi^2 r^2} , \tag{17}$$

$$\langle T_{uv} \rangle = -\frac{c}{96\pi^2 r^2 C^2} \left[C \partial_u \partial_v C - \partial_u C \partial_v C \right], \tag{18}$$

$$\langle T_{\Omega\Omega} \rangle = 0 , \tag{19}$$

where κ is the integration constant, which is related to the asymptotic condition. Then, it is straightforward to solve the semi-classical Einstein equation to obtain the expression near the Schwarzschild radius [30–33],

$$C(r_*) = e^{r_*/r_h} + \mathcal{O}(\alpha^2) \,, \tag{20}$$

$$r(r_*) = r_h + e^{r_*/r_h} - \alpha \left(\frac{1}{4r_h^2} - \kappa^2 \right) r_* + \mathcal{O}(\alpha^2) \,. \tag{21}$$

where $\alpha = \frac{cG_N}{12\pi}$ and r_* is the tortoise coordinate $r_* = \frac{1}{2}(v-u)$. The Hartle-Hawking vacuum is given by $\kappa = \frac{1}{2r_h}$ where the radius approaches to the Schwarzschild radius r_h in $r_* \to -\infty$. In the other static vacua, $\kappa \neq \frac{1}{2r_h}$, the radius diverges for $\kappa < \frac{1}{2r_h}$ or goes to zero for $\kappa > \frac{1}{2r_h}$ before $C(r_*)$ goes to zero. In either case, the geometry has no event horizon but naked singularity. Thus, two exteriors of the event horizon are disconnected in the other static vacua than the Hartle-Hawking vacuum.

Away from the Schwarzschild radius, the metric is approximated by the classical Schwarzschild solution (7), where

$$C(r_*) = 1 - \frac{r_h}{r(r_*)} \,, \tag{22}$$

$$r_* = r - r_h + r_h \log \left(\frac{r - r_h}{r_h} \right) \,. \tag{23}$$

5. Additivity conjecture in Schwarzschild spacetime

Now, we calculate the entanglement entropy in static vacua with $\kappa \neq \frac{1}{2r_h}$. As we have seen in the previous section, two exteriors are disconnected by taking quantum effects in the energy-momentum tensor into account. Then, the entanglement entropy of $R_1 \cup R_2$ is simply given by

$$S(R_1 \cup R_2) = S(R_1) + S(R_2) \,. \tag{24}$$

Thus, the additivity condition is satisfied.

Next, we calculate the entanglement entropy to see that the entanglement entropy in a vacuum with $\kappa \neq \frac{1}{2r_h}$ gives the minimum output entropy, or equivalently, is smaller than that in the Hartle-Hawking vacuum. In these vacua, the coordinates in the formula (9) should be chosen as

$$U = -\kappa^{-1}e^{-\kappa u} \,, \qquad\qquad V = \kappa^{-1}e^{\kappa v} \,, \tag{25}$$

where (u, v) coordinates are related to t and r_* as $v - t + r_*$ and $u - t - r_*$. In the configurations without islands, the entanglement entropy is calculated as [11]

$$S = \frac{\pi b^2}{G_N} + \frac{c}{12} \log \left(\frac{b - r_h}{\kappa^2 b} \right) . \tag{26}$$

This entanglement entropy becomes very large for very small κ, and then, a configuration with an island gives the dominant saddle point. The entanglement entropy for small κ is approximately given by [11]

$$S = \frac{\pi b^2}{G_N} + \frac{c}{12} \log \left(\frac{b - r_h}{b} \right) + \frac{c}{3} \log \left(\frac{48\pi r_h^3}{c\, G_N} \right) . \tag{27}$$

The entanglement entropy (26) is smaller than that of the Hartle-Hawking vacuum for $\kappa > \frac{1}{2r_h}$, and hence, gives the minimum output entropy. Therefore, the additivity conjecture is satisfied in the Schwarzschild spacetime if the other static vacua than the Hartle-Hawking vacuum are taken into consideration.

References

[1] P. Hayden and G. Penington, "Black hole microstates vs. the additivity conjectures," [arXiv:2012.07861 [hep-th]].

[2] A. S. Holevo, "The capacity of the quantum channel with general signal states," IEEE Transactions on Information Theory, 44 (1998) 269273.

[3] Benjamin Schumacher and Michael DWestmoreland, "Sending classical information via noisy quantum channels," Physical Review A 56 (1997) 131.

[4] C. King and M. B. Ruskai, "Minimal entropy of states emerging from noisy quantum channels," IEEE Transactions on Information Theory 47 (2001) 192209 [arXiv:quant-ph/9911079].

[5] A. A. Pomeransky, "Strong superadditivity of the entanglement of formation follows from its additivity," Physical Review A 68 (2003) 032317 [arXiv:quant-ph/0305056].

[6] K. M. R. Audenaert and S. L. Braunstein, "On strong superadditivity of the entanglement of formation," Communications in Mathematical Physics 246 (2004) 443452 [arXiv:quant-ph/0303045].

[7] K. Matsumoto, T. Shimono, and A. Winter, "Remarks on additivity of the Holevo channel capacity and of the entanglement of formation," Communications in Mathematical Physics, 246 (2004) 427442 [arXiv:quant-ph/0206148].

[8] Peter W. Shor, "Equivalence of additivity questions in quantum information theory," Communications in Mathematical Physics, 246 (2004) 453472.

[9] S. Ryu and T. Takayanagi, "Holographic derivation of entanglement entropy from AdS/CFT," Phys. Rev. Lett. **96** (2006), 181602 [arXiv:hep-th/0603001 [hep-th]].

[10] V. E. Hubeny, M. Rangamani and T. Takayanagi, "A Covariant holographic entanglement entropy proposal," JHEP **07** (2007), 062 [arXiv:0705.0016 [hep-th]].

[11] Y. Matsuo, "Entanglement entropy and vacuum states in Schwarzschild geometry," [arXiv:2110.13898 [hep-th]].

[12] T. Hartman and J. Maldacena, "Time Evolution of Entanglement Entropy from Black Hole Interiors," JHEP **05** (2013), 014 doi:10.1007/JHEP05(2013)014 [arXiv:1303.1080 [hep-th]].

[13] T. Faulkner, A. Lewkowycz and J. Maldacena, "Quantum corrections to holographic entanglement entropy," JHEP **11** (2013), 074 [arXiv:1307.2892 [hep-th]].

[14] A. Lewkowycz and J. Maldacena, "Generalized gravitational entropy," JHEP **08** (2013), 090 [arXiv:1304.4926 [hep-th]].

[15] N. Engelhardt and A. C. Wall, "Quantum Extremal Surfaces: Holographic Entanglement Entropy beyond the Classical Regime," JHEP **01** (2015), 073 [arXiv:1408.3203 [hep-th]].

[16] X. Dong, A. Lewkowycz and M. Rangamani, "Deriving covariant holographic entanglement," JHEP **11** (2016), 028 [arXiv:1607.07506 [hep-th]].

[17] X. Dong and A. Lewkowycz, "Entropy, Extremality, Euclidean Variations, and the Equations of Motion," JHEP **01** (2018), 081 [arXiv:1705.08453 [hep-th]].

[18] G. Penington, "Entanglement Wedge Reconstruction and the Information Paradox," JHEP **09** (2020), 002 [arXiv:1905.08255 [hep-th]].

[19] A. Almheiri, N. Engelhardt, D. Marolf and H. Maxfield, "The entropy of bulk quantum fields and the entanglement wedge of an evaporating black hole," JHEP **12** (2019), 063 [arXiv:1905.08762 [hep-th]].

[20] A. Almheiri, R. Mahajan, J. Maldacena and Y. Zhao, "The Page curve of Hawking radiation from semiclassical geometry," JHEP **03** (2020), 149 [arXiv:1908.10996 [hep-th]].

[21] A. Almheiri, R. Mahajan and J. Maldacena, "Islands outside the horizon," arXiv:1910.11077 [hep-th].

[22] G. Penington, S. H. Shenker, D. Stanford and Z. Yang, "Replica wormholes and the black hole interior," arXiv:1911.11977 [hep-th].

[23] A. Almheiri, T. Hartman, J. Maldacena, E. Shaghoulian and A. Tajdini, "Replica Wormholes and the Entropy of Hawking Radiation," JHEP **05** (2020), 013 [arXiv:1911.12333 [hep-th]].

[24] D. N. Page, "Information in black hole radiation," Phys. Rev. Lett. **71**, 3743 (1993) [hep-th/9306083].

[25] D. N. Page, "Time Dependence of Hawking Radiation Entropy," JCAP **1309**, 028 (2013) [arXiv:1301.4995 [hep-th]].

[26] K. Hashimoto, N. Iizuka and Y. Matsuo, "Islands in Schwarzschild black holes," JHEP **06** (2020), 085 [arXiv:2004.05863 [hep-th]].

[27] Y. Matsuo, "Islands and stretched horizon," JHEP **07** (2021), 051 [arXiv:2011.08814 [hep-th]].

[28] P. C. W. Davies, S. A. Fulling and W. G. Unruh, "Energy-momentum Tensor Near an Evaporating Black Hole," Phys. Rev. D **13**, 2720 (1976). doi:10.1103/PhysRevD.13.2720

[29] P. C. W. Davies and S. A. Fulling, "Radiation from a moving mirror in two-dimensional space-time conformal anomaly," Proc. Roy. Soc. Lond. A **348**, 393 (1976).

[30] A. Fabbri, S. Farese, J. Navarro-Salas, G. J. Olmo and H. Sanchis-Alepuz, "Semiclassical zero-temperature corrections to Schwarzschild spacetime and holography," Phys. Rev. D **73** (2006), 104023 [arXiv:hep-th/0512167 [hep-th]].

[31] A. Fabbri, S. Farese, J. Navarro-Salas, G. J. Olmo and H. Sanchis-Alepuz, J. Phys. Conf. Ser. **33** (2006), 457-462 [arXiv:hep-th/0512179 [hep-th]].

[32] P. M. Ho and Y. Matsuo, "Static Black Holes With Back Reaction From Vacuum Energy," Class. Quant. Grav. **35** (2018) no.6, 065012 [arXiv:1703.08662 [hep-th]].

[33] P. M. Ho and Y. Matsuo, "Static Black Hole and Vacuum Energy: Thin Shell and Incompressible Fluid," JHEP **03** (2018) 096 [arXiv:1710.10390 [hep-th]].

Holographic index calculation for Argyres-Douglas and Minahan-Nemeschansky theories

S. Murayama*

Department of Physics, Tokyo Institute of Technology,
Tokyo 152-8551, Japan
* E-mail: s.murayama@th.phys.titech.ac.jp

We calculate the superconformal indices of the Argyres-Douglas theories and Minahan-Nemeschansky theories realized on N coincident D3-branes in 7-brane backgrounds via the AdS/CFT correspondence. We include the finite N correction as "giant gravitons", which are D3-branes wrapping around 3-cycles. We take account of a single giant graviton for simplicity, and our method nicely reproduces known results and gives predictions for theories whose indices are unknown.

Keywords: Argyres-Douglas theories; Superconformal index; AdS/CFT correspondence.

1. Introduction

In this proceeding, we calculate the superconformal index of the Argyres-Douglas theories and Minahan-Nemeschansky theories using a brane realization in the type IIB superstring theory. We consider a brane system with N D3-branes and a 7-brane. Let x^0, x^1, x^2, and x^3 be the coordinates along D3-branes, and X, Y, and Z be the 3 complex coordinates of $\mathbb{C}^3$ transverse to the D3-branes. The D3-branes are located at $X = Y = Z = 0$. We also introduce a 7-brane with the worldvolume $Z = 0$ (Table 1).

Table 1. The brane setup.

	0	1	2	3	X	Y	Z
7-brane	✓	✓	✓	✓	✓	✓	
D3-branes	✓	✓	✓	✓			

There are 7 types of 7-branes with constant axiodilaton. Correspondingly, there are 7 types of 4-dimensional superconformal field theories, which we denote by $G[N]$. ($G = H_0, H_1, H_2, D_4, E_6, E_7, E_8$). This brane system

104

is quater BPS and $\mathcal{N} = 2$ supersymmetric theories are realized on the D3-brane worldvolume. $H_n[N]$ ($n = 0, 1, 2$) are Argyres-Douglas theories[1,2] and $E_n[N]$ ($n = 6, 7, 8$) are Minahan-Nemeschansky theories[3,4]. $D_4[N]$ is an SQCD with $Sp(N)$ gauge group. $H_n[N]$ and $E_n[N]$ are strongly coupled, and it is worth analyzing these SCFTs via AdS/CFT correspondence.

The 7-brane induces a deficit angle $\pi\alpha_G$ on Z-plane[5-8]. Table 2 shows α_G and the gauge symmetry on 7-brane for each G. From the viewpoint

Table 2. Theories corresponding to each dificit angle α_G.

G	H_0	H_1	H_2	D_4	E_6	E_7	E_8
α_G	1/3	1/2	2/3	1	4/3	3/2	5/3
gauge sym. on 7-brane	None	$SU(2)$	$SU(3)$	$SO(8)$	E_6	E_7	E_8

of 4-dimensional theories, the gauge symmetry on 7-brane becomes flavor symmetry. The rotational symmetry along transverse directions is $SO(4) \times SO(2) \sim SU(2)_R \times SU(2)_F \times U(1)_{R_Z}$, where $SU(2)_R \times U(1)_{R_Z}$ is R-symmetry, and $SU(2)_F$ is a flavor symmetry. Let R_X, R_Y, and R_Z be the generators rotating X, Y, and Z-plane, respectively. $U(1)_{R_Z}$ is generated by R_Z, and the Cartan generators of $SU(2)_R$ and $SU(2)_F$ are $\frac{1}{2}(R_X + R_Y)$ and $\frac{1}{2}(R_X - R_Y)$, respectively.

The near-horizon geometry of the brane system is $AdS_5 \times S^5_{\alpha_G}$, where $S^5_{\alpha_G}$ is a 5-sphere in XYZ space with deficit angle $\pi\alpha_G$ on Z-plane. The 7-brane wraps around the singular locus $Z = 0$, and the gauge symmetry G lives on the locus.

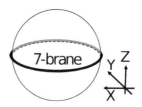

Fig. 1. 7-branes in $S^5_{\alpha_G}$ of $G[N]$.

The superconformal index is defined by[9]

$$\mathcal{I} = \text{tr}\left[e^{2\pi i(J+\bar{J})} q^{H+\bar{J}} y^{2J} u_x^{R_X} u_y^{R_Y} u_z^{R_Z} \prod_{i=1}^{\text{rank}G} x_i^{p_i} \right]. \quad (u_x u_y u_z = 1) \quad (1)$$

where H is the Hamiltonian, J and $\bar{J}$ are the angular momenta, and p_i are Cartan generators of G. It expresses the BPS spectrum concisely as a function of fugacities q, y, u_x, u_y, u_z and x_i. The factor $e^{2\pi i(J+\bar{J})}$ is $+1$ for bosonic states and -1 for fermionic states.

Table 3 shows the theories whose superconformal indices have already been calculated[10–14]. The method explained below can be used to calculate

Table 3. Theories whose superconformal index have been already calculated.

N	H_0	H_1	H_2	D_4	E_6	E_7	E_8
1	✓	✓	✓	✓	✓	✓	
2				✓			
3				✓			

the leading finite N correction to all $G[N]$.

In the following sections, we will briefly explain how to calculate the index on the AdS side and show a few results. For detailed explanations and more results, refer to the original paper[15] in collaboration with Y. Imamura.

2. Large N limit

Let us first consider the large N limit, which has already been analyzed in[16,17]. There are two contributions to the index (Figure 2).

(a) **Kaluza-Klein modes of the gravity multiplet in the bulk**
Closed strings give the gravity multiplet in the bulk. They are expanded into spherical harmonics in $S^5_{\alpha_G}$. Due to the deficit angle, the boundary condition associated with the angular coordinate on Z-plane should be appropriately modified[16,17]. The superconformal index can be obtained by summing up the contributions from all modes.

(b) **Kaluza-Klein modes of vector multiplets on the 7-brane**
Vector multiplets of gauge group G live on the 7-brane. We can expand them into spherical harmonics on the 7-brane worldvolume $S^3 \subset S^5_{\alpha_G}$ in a similar way to (a). The superconformal index can be obtained by summing up the contributions from all modes. In the D_4 case, the 7-brane is a stack of an O7-plane and 4 D7-branes, and the vector multiplet comes from open strings on the D7-branes.

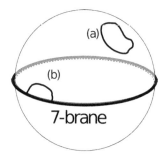

Fig. 2. The contributions in the large N limit. (a) The closed string in the bulk. (b) vector multiplets on the 7-brane. Open string description of (b) is justified only in the D_4 case.

3. Finite N corrections

For finite N, we include extra contributions from "giant gravitons" [18], which are the D3-branes extended in $S^5_{\alpha_G}$. We follow the prescription proposed in [19]. Namely, we consider D3-branes wrapped around three particular 3-cycles, $X = 0$, $Y = 0$, and $Z = 0$. It is expected the complete finite N index $\mathcal{I}_{G[N]}$ would be obtained if we sum up over all wrapping numbers (n_x, n_y, n_z) as

$$\mathcal{I}_{G[N]} = \mathcal{I}_{G[\infty]} \sum_{n_x, n_y, n_z} \mathcal{I}_{G[N]}^{(n_x, n_y, n_z)}, \qquad (2)$$

where $\mathcal{I}_{G[\infty]}$ is the large N index, and $\mathcal{I}_{G[N]}^{(n_x, n_y, n_z)}$ is the index from giant gravitons with specific wrapping numbers. It is difficult to calculate $\mathcal{I}_{G[N]}^{(n_x, n_y, n_z)}$ for general wrapping numbers, and here we take account of only leading corrections, $\mathcal{I}_{G[N]}^{(1,0,0)}$ and $\mathcal{I}_{G[N]}^{(0,1,0)}$.

$$\mathcal{I}_{G[N]} = \mathcal{I}_{G[\infty]}(1 + \mathcal{I}_{G[N]}^{(1,0,0)} + \mathcal{I}_{G[N]}^{(0,1,0)}) + \cdots . \qquad (3)$$

(Due to the deficit angle, the volumes of 3-cycles $X = 0$ and $Y = 0$ are smaller than that of $Z = 0$, and $\mathcal{I}_{G[N]}^{(0,0,1)}$ becomes subleading.) $\mathcal{I}_{G[N]}^{(1,0,0)}$ and $\mathcal{I}_{G[N]}^{(0,1,0)}$ are related to each other by the Weyl reflection $u_x \leftrightarrow u_y$.

We can split $\mathcal{I}_{G[N]}^{(1,0,0)}$ into 3 factors (Figure 3).

(c) **The classical factor from the maximal giant graviton**

The volume of the 3-cycle $X = 0$ is proportional to N, and a giant graviton wrapped around the cycle carries $H = R_X = N$. This corresponds to the factor $q^N u_x^N$.

(d) The vector multiplet on the giant graviton

$U(1)$ vector multiplet live on a giant graviton. Its fluctuations can be expanded into spherical harmonics. We sum up the contributions from them.

(e) The degrees of freedom along the intersection of the giant graviton and the 7-brane

In the D_4 case, open strings between the giant graviton and the D7-brane give chiral fermions living along the intersection. Their contribution to the index is the character of the basic representation of the $\hat{D}_4$ current algebra. For $G \neq D_4$, we cannot directly derive the contribution, and we assume that it is the character of the basic representation of the $\hat{G}$ current algebra.

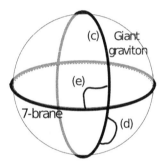

Fig. 3. Three ccontributins to the finite N correction. (c) The classical contribution. (d) The vector multiplet. (e) The intersection modes. Open string description of (e) is justified only in the D_4 case.

4. Results

Now, we can calculate the index by combining the five contributions (a)-(e). We will first show the known result for $H_0[1]^{10}$.

$$
\begin{aligned}
\mathcal{I}_{H_0[1]} =& 1 + u_z^{\frac{6}{5}} q^{\frac{6}{5}} - u_z^{\frac{1}{5}} \chi_1^J q^{\frac{17}{10}} + u_z^{-\frac{4}{5}} q^{\frac{11}{5}} \\
& + u_z^{\frac{12}{5}} q^{\frac{12}{5}} + u_z^{\frac{6}{5}} \chi_1^J q^{\frac{27}{10}} - u_z^{\frac{7}{5}} \chi_1^J q^{\frac{29}{10}} + \cdots,
\end{aligned} \tag{4}
$$

where χ_n^J and χ_n^F are the characters of the n-dimensional $SU(2)_J$ and $SU(2)_F$ representations, respectively.

Let us first compare this with the large N index without the giant graviton contribution included:

$$
\begin{aligned}
\mathcal{I}_{H_0[\infty]} =& 1 + u_z^{\frac{6}{5}} q^{\frac{6}{5}} - u_z^{\frac{1}{5}} \chi_1^J q^{\frac{17}{10}} + \underline{u_z^{-1} \chi_2^F q^2} + (u_z^{\frac{7}{10}} \chi_1^F + u_z^{-\frac{4}{5}}) q^{\frac{11}{5}} \\
& + 2 u_z^{\frac{12}{5}} q^{\frac{12}{5}} + (u_z^{\frac{6}{5}} - u_z^{-\frac{3}{10}} \chi_1^F) \chi_1^J q^{\frac{27}{10}} - 2 u_z^{\frac{7}{5}} \chi_1^J q^{\frac{29}{10}} + \cdots .
\end{aligned} \tag{5}
$$

We find the discrepancy between (5) and (4) at the order q^2 as shown by the underline.

Next, let us look at the result including the giant graviton contributions:

$$
\begin{aligned}
\mathcal{I}_{H_0[1]}^{\mathrm{AdS}} =& 1 + u_z^{\frac{6}{5}} q^{\frac{6}{5}} - u_z^{\frac{1}{5}} \chi_1^J q^{\frac{17}{10}} + u_z^{-\frac{4}{5}} q^{\frac{11}{5}} \\
& + u_z^{\frac{12}{5}} q^{\frac{12}{5}} + u_z^{\frac{6}{5}} \chi_1^J q^{\frac{27}{10}} \underline{+ u_z^{\frac{29}{5}} q^{\frac{14}{5}}} + \cdots .
\end{aligned} \tag{6}
$$

The discrepancy becomes of order $q^{\frac{14}{5}}$, which is higher than the previous result. This implies the contributions from giant gravitons nicely reproduces the finite N correction.

Finally, we show the result of $E_8[1]$

$$
\begin{aligned}
\mathcal{I}_{E_8[1]}^{\mathrm{AdS}} =& 1 + u_z^{-1} \chi_{\mathbf{248}}^{E_8} q^2 + (-1 - \chi_{\mathbf{248}}^{E_8}) q^3 + (u_z^{-1} + u_z^{-1} \chi_{\mathbf{248}}^{E_8}) \chi_1^J q^{\frac{7}{2}} \\
& + (u_z + u_z^2 \chi_{\mathbf{27000}}^{E_8}) q^4 + (-2 - \chi_{\mathbf{248}}^{E_8}) \chi_1^J q^{\frac{9}{2}} + \mathcal{O}(q^5),
\end{aligned} \tag{7}
$$

where $\chi_{\boldsymbol{n}}^{E_8}$ is the character of the n-dimensional irreducible E_8 representation. The expected error of our calculation is of order q^5, and all terms shown in (7) are expected to be correct. This has not been obtained by other methods.

Acknowledgments

We would like to thank Yosuke Imamura, Reona Arai, Shota Fujiwara, Tatsuya Mori, Takahiro Nishinaka, and Daisuke Yokoyama for valuable discussions and comments.

References

1. P. C. Argyres and M. R. Douglas, "New phenomena in SU(3) supersymmetric gauge theory," Nucl. Phys. B **448**, 93-126 (1995) doi:10.1016/0550-3213(95)00281-V [arXiv:hep-th/9505062 [hep-th]].
2. P. C. Argyres, M. R. Plesser, N. Seiberg and E. Witten, "New N=2 superconformal field theories in four-dimensions," Nucl. Phys. B **461**, 71-84 (1996) doi:10.1016/0550-3213(95)00671 0 [arXiv:hep-th/9511154 [hep-th]].

3. J. A. Minahan and D. Nemeschansky, "An N=2 superconformal fixed point with E(6) global symmetry," Nucl. Phys. B **482**, 142-152 (1996) doi:10.1016/S0550-3213(96)00552-4 [arXiv:hep-th/9608047 [hep-th]].

4. J. A. Minahan and D. Nemeschansky, "Superconformal fixed points with E(n) global symmetry," Nucl. Phys. B **489**, 24-46 (1997) doi:10.1016/S0550-3213(97)00039-4 [arXiv:hep-th/9610076 [hep-th]].

5. A. Sen, "F theory and orientifolds," Nucl. Phys. B **475**, 562-578 (1996) doi:10.1016/0550-3213(96)00347-1 [arXiv:hep-th/9605150 [hep-th]].

6. T. Banks, M. R. Douglas and N. Seiberg, "Probing F theory with branes," Phys. Lett. B **387**, 278-281 (1996) doi:10.1016/0370-2693(96)00808-8 [arXiv:hep-th/9605199 [hep-th]].

7. K. Dasgupta and S. Mukhi, "F theory at constant coupling," Phys. Lett. B **385**, 125-131 (1996) doi:10.1016/0370-2693(96)00875-1 [arXiv:hep-th/9606044 [hep-th]].

8. M. R. Douglas, D. A. Lowe and J. H. Schwarz, "Probing F theory with multiple branes," Phys. Lett. B **394**, 297-301 (1997) doi:10.1016/S0370-2693(97)00011-7 [arXiv:hep-th/9612062 [hep-th]].

9. J. Kinney, J. M. Maldacena, S. Minwalla and S. Raju, "An Index for 4 dimensional super conformal theories," Commun. Math. Phys. **275**, 209-254 (2007) doi:10.1007/s00220-007-0258-7 [arXiv:hep-th/0510251 [hep-th]].

10. K. Maruyoshi and J. Song, "Enhancement of Supersymmetry via Renormalization Group Flow and the Superconformal Index," Phys. Rev. Lett. **118**, no.15, 151602 (2017) doi:10.1103/PhysRevLett.118. 151602 [arXiv:1606.05632 [hep-th]].

11. K. Maruyoshi and J. Song, "$\mathcal{N} = 1$ deformations and RG flows of $\mathcal{N} = 2$ SCFTs," JHEP **02**, 075 (2017) doi:10.1007/JHEP02(2017)075 [arXiv:1607.04281 [hep-th]].

12. P. Agarwal, K. Maruyoshi and J. Song, "$\mathcal{N}$ =1 Deformations and RG flows of $\mathcal{N}$ =2 SCFTs, part II: non-principal deformations," JHEP **12**, 103 (2016) doi:10.1007/JHEP12(2016)103 [arXiv:1610.05311 [hep-th]].

13. A. Gadde, L. Rastelli, S. S. Razamat and W. Yan, "The Superconformal Index of the E_6 SCFT," JHEP **08**, 107 (2010) doi:10.1007/JHEP08(2010)107 [arXiv:1003.4244 [hep-th]].

14. P. Agarwal, K. Maruyoshi and J. Song, "A "Lagrangian" for the E_7 superconformal theory," JHEP **05**, 193 (2018) doi:10.1007/JHEP05(2018)193 [arXiv:1802.05268 [hep-th]].

15. Y. Imamura and S. Murayama, "Holographic index calculation for Argyres-Douglas and Minahan-Nemeschansky theories," [arXiv:2110.14897 [hep-th]].

16. A. Fayyazuddin and M. Spalinski, "Large N superconformal gauge theories and supergravity orientifolds," Nucl. Phys. B **535**, 219-232 (1998) doi:10.1016/S0550-3213(98)00545-8 [arXiv:hep-th/9805096 [hep-th]].

17. O. Aharony, A. Fayyazuddin and J. M. Maldacena, "The Large N limit of N=2, N=1 field theories from three-branes in F theory," JHEP **07**, 013 (1998) doi:10.1088/1126-6708/1998/07/013 [arXiv:hep-th/9806159 [hep-th]].

18. J. McGreevy, L. Susskind and N. Toumbas, "Invasion of the giant gravitons from Anti-de Sitter space," JHEP **06**, 008 (2000) doi:10.1088/1126-6708/2000/06/008 [arXiv:hep-th/0003075 [hep-th]].

19. R. Arai and Y. Imamura, "Finite N Corrections to the Superconformal Index of S-fold Theories," PTEP **2019**, no.8, 083B04 (2019) doi:10.1093/ptep/ptz088 [arXiv:1904.09776 [hep-th]].

Topological defect junctions in 4-dimensional pure $\mathbb{Z}_2$ lattice gauge theory

Y. Nagoya*

Department of Physics, Graduate School of Science, Osaka University,
Machikane-yama-cho 1-1, Toyonaka 560-0043, Osaka, Japan
* E-mail: y_nagoya@het.phys.sci.osaka-u.ac.jp

We explore topological defects and their properties in 4-dimensional pure $\mathbb{Z}_2$ lattice gauge theory. This theory has the Kramers-Wannier-Wegner(KWW) duality. Duality defects associated with the KWW duality are constructed and shown to be non-invertible topological defects. In this paper, we explore the crossing relations including the duality defects. We construct 1-form $\mathbb{Z}_2$ center symmetry defects and defect junctions. Crossing relations are derived from these defects and defect junctions. We also calculate some expectation values of topological defects by crossing relations.

Keywords: Generalized global symmetries, Non-invertible symmetries

1. Introduction

The concept of symmetry in quantum field theories is very important. There are many applications to the non-perturbative analysis of quantum fields theories. In recent years, there has been progress in the generalization of the notion of symmetries. The important notion of generalizations is topological defects. One of the generalizations is so-called non-invertible symmetries. There are a lot of studies of non-invertible symmetries in 2-dimensions[1–14]. Non-invertible symmetries in 2-dimensions are relatively understood than in higher dimensions. There are several studies of non-invertible symmetries in higher dimensions[15–22].

In this paper, we study 4-dimensional $\mathbb{Z}_2$ pure lattice gauge theory. There is a duality discovered by Wagner[23] in this model. This duality is similar to so-called Kramers-Wannier duality[24 25], therefore we call this duality "KWW duality". Another important property is a 1-form $\mathbb{Z}_2$ center symmetry[26]. The charged objects are Wilson loops.

In this paper, we aim to investigate the crossing relations among topological defects in 4-dimensional $\mathbb{Z}_2$ lattice gauge theory based on this work[27] which is in collaboration with Masataka Koide and Satoshi Yamaguchi.

Crossing relations are local relations between topological defects. These relations determine the algebra of symmetry structure. The crossing relations are not closed within only duality defects. For the crossing relations, we need $\mathbb{Z}_2$ symmetry defects and junctions. Defect junctions occur when two types of defects meet. We determine the weight of junctions by junction commutation relations and find the crossing relations by these defects and defect junctions. Related examples are discovered by the works[28,29].

This paper is organized as follows. In Sec. 2, we explain our setup and KWW duality defects. In Sec. 3, we construct $\mathbb{Z}_2$ symmetry defects and defect junctions. In Sec. 4, we explain crossing relations and calculate some expectation values of KWW duality defects.

2. 4-dimensional $\mathbb{Z}_2$ pure lattice gauge theory

In this section, we explain our setup of 4-dimensional $\mathbb{Z}_2$ lattice gauge theory. We prepare two kinds of 4-dimensional cubic lattices. One lattice is an active lattice $\Lambda := \{(x_1, x_2, x_3, x_4) | x_1, x_2, x_3, x_4 \in 2\mathbb{Z}\}$. Another lattice is an inactive lattice $\hat{\Lambda} := \{(x_1, x_2, x_3, x_4) | x_1, x_2, x_3, x_4 \in 2\mathbb{Z} + 1\}$. A schematic picture of these lattices is shown in Fig. 1

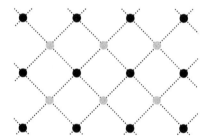

Fig. 1. A schematic picture of our lattices. Even though this figure is drawn in 2-dimensional, the lattices are 4-dimensional. Black dots represent a lattice Λ. Blue dots represent a lattice $\hat{\Lambda}$. These lattices are dual to each other.

We assign the link variables $U_m = (-1)^{a_m} = \pm 1$ to each link on Λ. We also call $a_m (= 0, 1)$ a link variable. We don't assign the link variables to each link on $\hat{\Lambda}$. There is a one-to-one correspondence between 16-cells and plaquettes on Λ. We regard the 16-cells as fundamental components of the total lattice. We assign the Boltzmann weight to the 16-cell with link variables $a_k = 0, 1 (k = 1, 2, 3, 4)$ and real parameter K as

$$W(a_1, a_2, a_3, a_4) = \exp(K(-1)^{(a_1 + a_2 + a_3 + a_4)}), \tag{1}$$

We define the partition function as

$$Z = \sum_{\{a\}} (\prod_{\substack{\text{active} \\ \text{sites}}} s)(\prod_{\substack{\text{active} \\ \text{links}}} l)(\prod_{\substack{\text{inactive} \\ \text{sites}}} \bar{s})(\prod_{\substack{\text{inactive} \\ \text{links}}} \bar{l}) \prod_{i \in C} W(a_{j_1(i)}, a_{j_2(i)}, a_{j_3(i)}, a_{j_4(i)}),$$

(2)

where C is the set of all 16-cells and $j_1(i)$, $j_2(i)$, $j_3(i)$, $j_4(i)$ are links of a plaquette i in Λ. The partition function sums over all configurations of link variables. s and l are the weights of the site and the link on Λ, respectively. $\bar{s}$ and $\bar{l}$ are the weights of the site and the link on $\hat{\Lambda}$, respectively.

We construct KWW duality defects on the codimension 1 surfaces. Weights of links, sites and KWW duality defects, and K are determined by commutation relations of KWW duality defects. Commutation relations are requirements that KWW duality defects can be smoothly deformed. Because we expect the KWW duality defects are topological defects, we require these commutation relations. The nontrivial solution is

$$s = \frac{1}{\sqrt{2}}, \quad l = \frac{1}{\sqrt{2}}, \quad \bar{s} = 1, \quad \bar{l} = 1, \quad K = \frac{1}{2}\log(1 + \sqrt{2}). \quad (3)$$

The solution of KWW defects has the following properties. Wilson loops have nontrivial action of KWW duality defects. When the duality defect acts to a Wilson loop, a 't Hooft loop appears where the Wilson loop was. The most important property is that KWW duality defects are non-invertible.

3. $\mathbb{Z}_2$ 1-form symmetry defects and defect junctions

In this section, we construct $\mathbb{Z}_2$ 1-form symmetry defects and defect junctions. There are 1-form $\mathbb{Z}_2$ center symmetries[26] in this model. We need to construct 1-form $\mathbb{Z}_2$ center symmetry defects as codimension 2 objects. We consider that $\mathbb{Z}_2$ symmetry defects are supported on triangles which are formed by a link in $\hat{\Lambda}$ and the midpoint of the adjusted link in Λ. We deformed these triangles to double the link in $\hat{\Lambda}$ and the midpoint of the adjust link in Λ on these triangles. On the other hand, the sites at both ends in Λ of these triangles are not doubled. These deformations are shown in Fig. 2.

We assign weights to $Z_2(b,c) = \sigma^x_{b,c} = (1 - \delta_{b,c})$ to each building block of $\mathbb{Z}_2$ symmetry defects. $b, c = 0, 1$ are link variables in the building block. We also assign the weight $z = \sqrt{2}$ to each pair of each link on Λ. $\mathbb{Z}_2$ symmetry defects assigned those weights are topological and invertible. Topological means that $\mathbb{Z}_2$ symmetry defects can be smoothly deformed on octagons

114

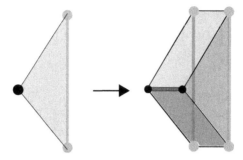

Fig. 2. A schematic picture of the $\mathbb{Z}_2$ symmetry defect. Blue dots and links represent sites and links in $\hat{\Lambda}$. Black dots are the midpoints of links in Λ. We assign weights a triangular prism and a red line. We deform triangles to double the links and sites on triangles.

surrounded by triangles. We can show these defects are invertible from the fact that the S^2 expectation value of a $\mathbb{Z}_2$ symmetry defect is 1. When the $\mathbb{Z}_2$ symmetry defects act on a Wilson loop, the sign flips. Therefore, we can consider the constructed $\mathbb{Z}_2$ symmetry defects are actually the symmetry defects associated with a 1-form $\mathbb{Z}_2$ center symmetry.

We consider configurations of defects where KWW duality defects and $\mathbb{Z}_2$ symmetry defects meet. Junctions occur in such configurations. There are two types of junctions depending on whether a sharing link is in Λ or $\hat{\Lambda}$. Fig. 3 and Fig. 4 show schematic pictures of two types of junctions. We denote each weight of junctions $J(a)$ and $\tilde{J}(b,c)$. $a, b, c = 0, 1$ are link variables in these junctions.

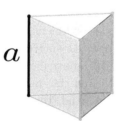

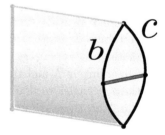

Fig. 3. A schematic picture of a junction. We call its weight $J(a)$. a is a link variable in this junction.

Fig. 4. A schematic picture of a junction. We call its weight $\tilde{J}(b,c)$. b, c are link variables in this junction.

These weights are determined by junction commutation relations. Junction commutation relations are a requirement that $\mathbb{Z}_2$ symmetry defects can be smoothly deformed along with KWW duality defects as shown in Fig. 5. We expect there are topological junctions and for this reason, we require these junction commutation relations. The nontrivial solution is

$$J(a) = (-1)^a, \quad \tilde{J}(b,c) = \sigma_{b,c}^x. \tag{4}$$

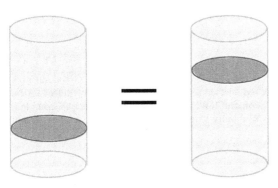

Fig. 5. A schematic picture of the junction commutation relations. Green surfaces represent KWW duality defects. Red surfaces represent a $\mathbb{Z}_2$ symmetry defect. This relation is a requirement that the junctions can be smoothly deformed.

4. Crossing relations

In this section, we explain crossing relations between KWW duality defects and $\mathbb{Z}_2$ symmetry defects. Crossing relations can be explicitly calculated by weights of defects and junctions. We find three types of crossing relations.

The first crossing relations are as shown in Fig. 6. $\mathbb{Z}_2$ symmetry defects that have the end to KWW duality defects can be removed when the boundary of a $\mathbb{Z}_2$ symmetry defect on KWW duality defects is homologically trivial.

The Second crossing relations are as shown in Fig. 7. We call these relations solid torus equations. We consider a decomposition of S^3 into two 3-dimensional solid tori. We call these solid tori V_1 and V_2. We place a KWW duality defect on V_1 and this is the left-hand side of Fig. 7 . The first term of the right-hand side of Fig. 7 is the configuration of KWW duality defect on V_2. The second term of the right-hand side of Fig. 7 is the configuration of KWW duality defect on V_2 with a $\mathbb{Z}_2$ symmetry defect on D_2 whose boundary is a non-trivial cycle on V_2.

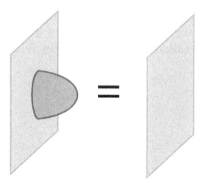

Fig. 6. A schematic picture of the crossing relations. Green surfaces represent KWW duality defects. A red surface represents a $\mathbb{Z}_2$ symmetry defect. A boundary of the $\mathbb{Z}_2$ symmetry defect is on the KWW duality defect. When the boundary of $\mathbb{Z}_2$ symmetry defects on the KWW duality defect is homologically trivial, the $\mathbb{Z}_2$ symmetry defect can be removed.

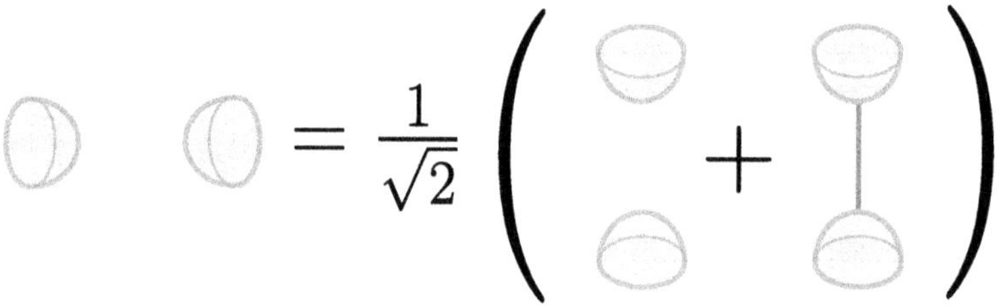

$$
\bigcirc \quad \bigcirc = \frac{1}{\sqrt{2}} \left(\quad + \quad \right)
$$

Fig. 7. A schematic picture of solid torus equations. Green surfaces represent KWW duality defects. A red line represents a $\mathbb{Z}_2$ symmetry defect. The left-hand side is a configuration of a KWW duality defect on V_1. The first tern of the left-hand side is a configuration of a KWW duality defect on V_2 and the second term of it is a configuration of a KWW duality defect on V_2 with a $\mathbb{Z}_2$ symmetry defect on D_2 whose boundary is a non-trivial cycle on V_2.

The third crossing relations are as shown in Fig. 8. We consider the decomposition of S^3 into $S^0 \times D^3$ and $D^1 \times S^2$. We consider the configuration of KWW duality defects on $S^0 \times D^3$ and $D^1 \times S^2$. Expectation values of these configurations are $1/\sqrt{2}$ times different.

We can calculate some expectation values of a configuration of a KWW duality defect by crossing relations.

One example is an expectation value of a KWW duality defect on S^3. Let us consider two S^3 expectation values of KWW duality defects. Each S^3 has D^3. Therefore we can use a crossing relation of Fig. 8. After using a crossing relation of Fig. 8, we can find the expectation value of S^3 is $1/\sqrt{2}$.

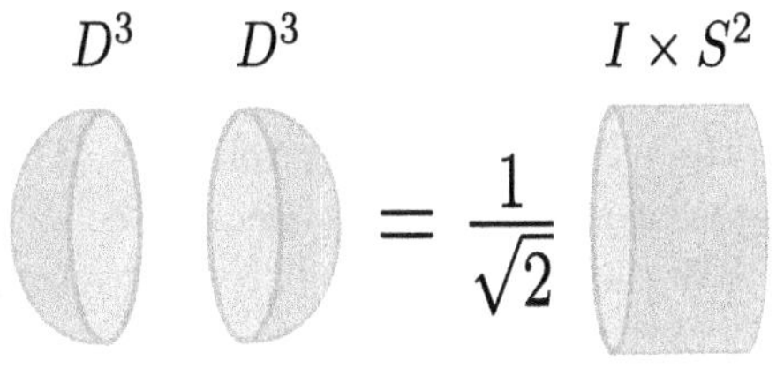

$$D^3 \qquad D^3 \qquad\qquad I \times S^2$$

$$= \frac{1}{\sqrt{2}}$$

Fig. 8.   A schematic picture of the crossing relation including $S^0 \times D^3$.

This expectation value means KWW duality defects are non-invertible.

Another example is an expectation value of a KWW duality defect on $S^2 \times S^1$. We can calculate this expectation value by using crossing relations of Fig. 6 and Fig. 7. A schematic picture of a calculation is as shown in Fig. 9. $S^2 \times S^1$ contains $D^2 \times S^1$. This is a solid torus. Therefore, we can use a solid torus equation of Fig. 7 and we use this in the first equality of Fig. 9. A $\mathbb{Z}_2$ symmetry defect on D^2 and its boundary is on a KWW duality defect on S^3. We use crossing relations of Fig. 6 to this term. We can see that the $S^2 \times S^1$ expectation value is 1 by using the result of an S^3 expectation value of a KWW duality defect.

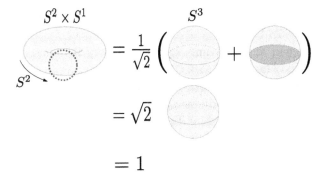

$$S^2 \times S^1 \qquad\qquad\qquad S^3$$

$$= \frac{1}{\sqrt{2}} \left(\quad + \quad \right)$$

$$= \sqrt{2}$$

$$= 1$$

Fig. 9.   A schematic picture of a calculation of an expectation value of a KWW duality defect on $S^2 \times S^1$. In the first equality, we use the solid torus equation of Fig. 7. In the second equality, we use a crossing relation of Fig. 6. In the third equality, we use the result that the S^3 expectation value of a KWW duality defect is $1/\sqrt{2}$.

5. Conclusions

In this paper, we construct topological defects in 4-dimensional pure $\mathbb{Z}_2$ lattice gauge theory. There are KWW duality defects and 1-form $\mathbb{Z}_2$ symmetry defects. KWW duality defects are codimension1 non-invertible topological defects. We construct 1-form $\mathbb{Z}_2$ symmetry defects and defect junctions. Finally, we explain three types of crossing relations. These relations can be calculated by weights of junctions and defects. We calculate some expectation values of KWW duality defects by crossing relations.

References

1. A. Feiguin, S. Trebst, A. W. W. Ludwig, M. Troyer, A. Kitaev, Z. Wang and M. H. Freedman, Interacting anyons in topological quantum liquids: The golden chain, *Phys. Rev. Lett.* **98**, p. 160409 (2007).
2. N. Carqueville and I. Runkel, Orbifold completion of defect bicategories, *Quantum Topol.* **7**, p. 203 (2016).
3. I. Brunner, N. Carqueville and D. Plencner, A quick guide to defect orbifolds, *Proc. Symp. Pure Math.* **88**, 231 (2014).
4. L. Bhardwaj and Y. Tachikawa, On finite symmetries and their gauging in two dimensions, *JHEP* **03**, p. 189 (2018).
5. C.-M. Chang, Y.-H. Lin, S.-H. Shao, Y. Wang and X. Yin, Topological Defect Lines and Renormalization Group Flows in Two Dimensions, *JHEP* **01**, p. 026 (2019).
6. Y.-H. Lin and S.-H. Shao, Duality Defect of the Monster CFT, *J. Phys. A* **54**, p. 065201 (2021).
7. R. Thorngren and Y. Wang, Fusion Category Symmetry I: Anomaly In-Flow and Gapped Phases (12 2019).
8. Z. Komargodski, K. Ohmori, K. Roumpedakis and S. Seifnashri, Symmetries and strings of adjoint QCD_2, *JHEP* **03**, p. 103 (2021).
9. T.-C. Huang and Y.-H. Lin, Topological Field Theory with Haagerup Symmetry (2 2021).
10. K. Inamura, Topological field theories and symmetry protected topological phases with fusion category symmetries, *JHEP* **05**, p. 204 (2021).
11. R. Thorngren and Y. Wang, Fusion Category Symmetry II: Categoriosities at $c = 1$ and Beyond (6 2021).
12. E. Sharpe, Topological operators, noninvertible symmetries and decomposition (8 2021).
13. I. M. Burbano, J. Kulp and J. Neuser, Duality Defects in E_8 (12 2021).
14. K. Inamura, On lattice models of gapped phases with fusion category symmetries (10 2021).
15. W. Ji and X.-G. Wen, Categorical symmetry and noninvertible

anomaly in symmetry-breaking and topological phase transitions, *Phys. Rev. Res.* **2**, p. 033417 (2020).

16. L. Kong, T. Lan, X.-G. Wen, Z.-H. Zhang and H. Zheng, Algebraic higher symmetry and categorical symmetry – a holographic and entanglement view of symmetry, *Phys. Rev. Res.* **2**, p. 043086 (2020).

17. T. Rudelius and S.-H. Shao, Topological Operators and Completeness of Spectrum in Discrete Gauge Theories, *JHEP* **12**, p. 172 (2020).

18. T. Johnson-Freyd, (3+1)D topological orders with only a $\mathbb{Z}_2$-charged particle (11 2020).

19. B. Heidenreich, J. Mcnamara, M. Montero, M. Reece, T. Rudelius and I. Valenzuela, Non-Invertible Global Symmetries and Completeness of the Spectrum (4 2021).

20. L. Bhardwaj, S. Giacomelli, M. Hübner and S. Schäfer-Nameki, Relative Defects in Relative Theories: Trapped Higher-Form Symmetries and Irregular Punctures in Class S (12 2021).

21. M. Nguyen, Y. Tanizaki and M. Ünsal, Semi-Abelian gauge theories, non-invertible symmetries, and string tensions beyond N-ality, *JHEP* **03**, p. 238 (2021).

22. M. Nguyen, Y. Tanizaki and M. Ünsal, Non-invertible 1-form symmetry and Casimir scaling in 2d Yang-Mills theory (4 2021).

23. F. J. Wegner, Duality in Generalized Ising Models and Phase Transitions Without Local Order Parameters, *J. Math. Phys.* **12**, 2259 (1971).

24. H. A. Kramers and G. H. Wannier, Statistics of the two-dimensional ferromagnet. Part I, *Phys. Rev.* **60**, p. 252 (1941).

25. H. A. Kramers and G. H. Wannier, Statistics of the two-dimensional ferromagnet. Part II, *Phys. Rev.* **60**, p. 263 (1941).

26. D. Gaiotto, A. Kapustin, N. Seiberg and B. Willett, Generalized Global Symmetries, *JHEP* **02**, p. 172 (2015).

27. M. Koide, Y. Nagoya and S. Yamaguchi, Non-invertible topological defects in 4-dimensional $\mathbb{Z}_2$ pure lattice gauge theory (9 2021).

28. J. Kaidi, K. Ohmori and Y. Zheng, Kramers-Wannier-like duality defects in (3+1)d gauge theories (11 2021).

29. Y. Choi, C. Cordova, P.-S. Hsin, H. T. Lam and S.-H. Shao, Non-Invertible Duality Defects in 3+1 Dimensions (11 2021).

Regge conformal blocks from the Rindler-AdS black hole and the pole-skipping phenomena

Mitsuhiro Nishida

School of Physics and Chemistry, Gwangju Institute of Science and Technology,
123 Cheomdan-gwagiro, Gwangju 61005, Korea
E-mail: mnishida@gist.ac.kr

We construct half-geodesic Witten diagrams with four external scalar fields in the Rindler-AdS black hole and show that their late-time behaviors agree with the Regge limit of conformal blocks. We also argue connection with pole-skipping phenomena by demonstrating that the near-horizon analysis in the Rindler-AdS black hole can determine the Regge behaviors of conformal blocks. This proceeding is based on a collaboration with Keun-Young Kim and Kyung-Sun Lee.

Keywords: AdS/CFT correspondence; Black hole; Conformal block

1. Introduction and summary

The Regge limit of conformal four-point functions is a well-studied subject for consistency of quantum field theories. Out-of-time-order correlation functions (OTOCs) in conformal field theories (CFTs) are related to the Regge limit, and it has been proposed that exponential behaviors in four-point OTOCs are bounded by the consistency[1]. It has also been proposed that the exponential behaviors in theories with Einstein gravity duals can be calculated from pole-skipping points in retarded Green's functions of the energy density[2,3]. Here, the pole-skipping points are defined by intersections between poles and zeros of momentum Green's functions.

In our paper[4], from the viewpoint of holography, we study a relation between the Regge limit of conformal blocks and the pole-skipping points in CFTs on Rindler spacetime. First, we construct half-geodesic Witten diagrams in the two-sided Rindler-AdS black hole and demonstrate that their late-time behaviors agree with the Regge limit of conformal blocks. Second, we show that the near-horizon analysis in the Rindler-AdS black hole for the pole-skipping points can capture the Regge behaviors of conformal blocks. Our result is a generalization of the original proposal regarding the pole-skipping points.

2. Half-geodesic Witten diagrams and the Regge limit of conformal blocks

We define scattering amplitudes of spin-ℓ exchange half-geodesic Witten diagrams in the Rindler-AdS black hole as

$$
\begin{aligned}
\mathcal{W}^{\mathcal{R}}_{\Delta,\ell} := \int_{\gamma^R_W} d\lambda \int_{\gamma^L_V} d\lambda' & G_{b\partial}\left(Y(\lambda), W_L; \Delta_W\right) G_{b\partial}\left(Y(\lambda), W_R; \Delta_W\right) \\
& \times G_{bb}\left(Y(\lambda), Y(\lambda'); \frac{dY(\lambda)}{d\lambda}, \frac{dY(\lambda')}{d\lambda'}; \Delta, \ell\right) \\
& \times G_{b\partial}\left(Y(\lambda'), V_L; \Delta_V\right) G_{b\partial}\left(Y(\lambda'), V_R; \Delta_V\right),
\end{aligned}
\tag{1}
$$

where $G_{b\partial}$ is a scalar bulk-to-boundary propagator, and G_{bb} is a spin-ℓ bulk-to-bulk propagator. Instead of the entire black hole spacetime, we integrate over two half-geodesics γ^R_W and γ^L_V between boundary points and centers of Penrose diagrams. After some approximations and calculations, we evaluate (1) in a late-time limit:

$$
\begin{aligned}
\mathcal{W}^{\mathcal{R}}_{\Delta,\ell} \simeq & \frac{\left(\mathcal{C}_{\Delta_W,0}\mathcal{C}_{\Delta_V,0}\right)^2 \mathcal{C}_{\Delta,0}}{2^{2(\Delta_W+\Delta_V)+\ell}(\Delta-1)} \log\left(\frac{1}{\epsilon}\right) \\
& \times e^{(\ell-1)t_R-(\Delta-1)\mathbf{d}} \,_2F_1\left(\Delta-1, \frac{d}{2}-1, \Delta+1-\frac{d}{2}; e^{-2\mathbf{d}}\right),
\end{aligned}
\tag{2}
$$

which agrees with the Regge limit of conformal blocks[5,6]. Here, t_R and $\mathbf{d}$ are differences of time and space between two operators in four-point OTOCs.

3. Near-horizon analysis in the Rindler-AdS black hole

Let us consider an ansatz of symmetric traceless spin-ℓ fields

$$
h_{v...v\mu}(v, r, \mathbf{x}) = e^{-i\omega v} \sum_{j=0}^{\infty} (r-1)^j h^{(j)}_{v...v\mu}(\mathbf{x}),
\tag{3}
$$

where v is a coordinate in the incoming Eddington-Finkelstein coordinates of the Rindler-AdS black hole. By substituting this expansion into the classical equations of free spin-ℓ fields, we obtain

$$
\begin{aligned}
0 &= (\nabla_\mu \nabla^\mu - \Delta(\Delta - d) + \ell) h_{v...v}(v, r, \mathbf{x}) \\
&= e^{-i\omega v} \big[\Box_\mathbb{H} - (\ell + i\omega)(d - 1) - \Delta(\Delta - d) \big] h^{(0)}_{v...v}(\mathbf{x}) \\
&\quad -2e^{-i\omega v} \big[i\omega + (\ell - 1) \big] h^{(1)}_{v...v}(\mathbf{x}) - 2\ell e^{-i\omega v} \big[i\omega + (\ell - 1) \big] h^{(0)}_{v...vr}(\mathbf{x}) + \cdots .
\end{aligned}
\tag{4}
$$

The near-horizon analysis is a holographic method to search pole-skipping points by imposing that coefficients in the classical equations are zero[7]. From the near-horizon analysis of (4), we obtain conditions for the pole-skipping points:

$$
-i\omega = \ell - 1, \qquad \big[\Box_\mathbb{H} - (\Delta - 1)(\Delta - d + 1) \big] h^{(0)}_{v...v} = 0 .
\tag{5}
$$

These conditions are consistent with the Regge behaviors of conformal blocks (2) with respect to t_R and $\mathbf{d}$[5,6], and therefore the pole-skipping points in CFTs on Rindler spacetime are related to the Regge limit of conformal blocks.

References

1. J. Maldacena, S. H. Shenker and D. Stanford, A bound on chaos, *JHEP* **08**, p. 106 (2016).
2. S. Grozdanov, K. Schalm and V. Scopelliti, Black hole scrambling from hydrodynamics, *Phys. Rev. Lett.* **120**, p. 231601 (2018).
3. M. Blake, H. Lee and H. Liu, A quantum hydrodynamical description for scrambling and many-body chaos, *JHEP* **10**, p. 127 (2018).
4. K.-Y. Kim, K.-S. Lee and M. Nishida, Regge conformal blocks from the Rindler-AdS black hole and the pole-skipping phenomena, *JHEP* **11**, p. 020 (2021).
5. L. Cornalba, M. S. Costa, J. Penedones and R. Schiappa, Eikonal Approximation in AdS/CFT: Conformal Partial Waves and Finite N Four-Point Functions, *Nucl. Phys. B* **767**, 327 (2007).
6. L. Cornalba, Eikonal methods in AdS/CFT: Regge theory and multi-reggeon exchange (10 2007).
7. M. Blake, R. A. Davison, S. Grozdanov and H. Liu, Many-body chaos and energy dynamics in holography, *JHEP* **10**, p. 035 (2018).

Entanglement entropy in interacting field theories

Katsuta Sakai*

KEK Theory Center, Institute of Particle and Nuclear Studies, High Energy Accelerator Research Organization (KEK), Oho 1-1, Tsukuba, Ibaraki 305-0801, Japan
* E-mail: sakaika@post.kek.jp

Entanglement entropy (EE) in field theory is a measure for quantum entanglement between spatially separated regions. While there are a lot of studies on EE in CFTs and free theories, EE in general interacting field theories requires further investigation. It is of very interest in order to relate the effect of the entanglement with low-energy physics. In this talk, we introduce our study on EE in interacting field theories with a subregion of a half-space. There, specific contributions to EE can be expressed in terms of renormalized correlation functions of operators. The contributions are expected to be dominant when we discuss low-energy effective theories.

Keywords: Entanglement entropy; Field theory; Correlation functions

1. Introduction

In the last few decades, there has been much investigation on the entanglement entropy (EE) , which is one of standard measures for the bipartite quantum entanglement between subsystems. In particular in field theory, while two causally separated regions are irrelevant at the classical level, the quantum entanglement between them can bring some nontrivial influence on observations made in the subsystem. EE can capture such a nontrivial quantum effect, and has been studied in many contexts: quantum phase transition, information paradox of the blackhole, holography, and so on.

So far, EE has been widely discussed in CFTs[1-3] or perturbations from them[4-6], and in free theories[7-13]. On the other hand, EE in interacting field theories which are far from CFTs is rather less understood. There are conceptional and technical difficulties. EE measures the entanglement with a physical cutoff scale, and we are often interested in not the entire EE but its "universal" part. In an interacting field theory, we are to deal with divergent radiative corrections by renormalization, and the physical quantities are described without explicit appearance of the cutoff scale. Here, a natural question is how the universal part of EE is related to the renormalized physical quantities. It is of course crucial to directly investigate the physical implication of EE to realistic observations. Despite its importance, not so many studies have been made on it because it is very difficult due to the lack of any useful symmetries or simplifications. On the other hand, it was reported in Ref. 14 that a part of EE in a scalar field theory is expressed with the renormalized mass parameter in a specific setup.

In this talk, we present our series of works on EE in interacting field theories [15–17], where we apply a technique called the orbifold technique [18,19]. It was originally proposed to evaluate EE in free field theory in the case of A being the flat half-space, and we apply it to interacting field theories. There, by investigating the general structure of Feynman bubble diagrams, we check that the area-law of EE holds at all order. Then we extract an essential part of EE which comes from two-point correlation functions of various operators. The two-point functions naturally contains the radiative corrections and is described by renormalized parameters. As a result, we associate a dominant part of EE with renormalized correlation functions of fundamental and composite operators. We show our analysis with an example of an interacting massive scalar field theory, while the basic idea is applicable to more general theories, as long as we consider the half-space subregion.

In Section 2, we explain our application of the orbifold technique to interacting field theories, and derive the area-law of EE to all order. We describe in Section 3 our main result, where EE is associated with the renormalized correlation functions. Section 4 is devoted to the summary and outlook.

2. Orbifold technique in interacting field theories

In field theory, a state is defined on a Cauchy surface, namely a spacelike hypersurface. Consider separating the surface into two subregions: A and $\bar{A}$. Those who can make an observation only in A find him/herself in the state $\rho_A = \mathrm{Tr}_{\bar{A}}\,\rho_{\mathrm{tot}}$, where ρ_{tot} is the density matrix for the total system. The corresponding EE is defined as the von Neumann entropy of ρ_A:

$$S_A = -\mathrm{Tr}_A\left(\rho_A \log \rho_A\right). \tag{1}$$

A standard way to calculate it is the replica trick [7], where we consider an n-fold covering. It consists of n replicas of the Euclidean system sewed up together at the subregion A. EE is then obtained by calculating the free

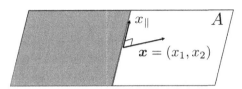

Fig. 1. Our setup to investigate EE. The subregion A is a half-space, whose boundary is expressed with a red line. The coordinates perpendicular and parallel to the boundary are x and $x_{\|}$, respectively.

energy of the theory on the n-fold, taking the derivative of it with respect to n, and taking the limit $n \to 1$.

In particular, when we focus on the case where A is a half-space in d-dimensional flat space (see Fig. 1), we can employ a variation of the replica trick called the orbifold technique. In this technique, we take an analytical continuation on n to $1/M$ with a integer M. Then, EE is computed as

$$S_A = -\frac{\partial \left(M F^{(M)} \right)}{\partial M}\bigg|_{M \to 1}, \tag{2}$$

where $F^{(M)}$ is the free energy of the theory on an orbifold $\mathbb{R}^2/\mathbb{Z}_M$. In the following, we represent the two-dimensional coordinates perpendicular to the boundary as $x = (x_1, x_2)$, and the others parallel to the boundary as $x_{\|}$.

The advantage of introducing the orbifold is that we can formulate the theory on it almost in the same manner as on the ordinary flat space. The only difference is that we introduce a projection operator, that is a symmetric summation of a rotation operator:

$$\hat{P} = \frac{1}{M} \sum_{m=0}^{M-1} \hat{g}^m, \quad \hat{g} : \begin{pmatrix} x_1 \\ x_2 \end{pmatrix} \mapsto \begin{pmatrix} \cos \frac{2\pi}{M} & -\sin \frac{2\pi}{M} \\ \sin \frac{2\pi}{M} & \cos \frac{2\pi}{M} \end{pmatrix} \begin{pmatrix} x_1 \\ x_2 \end{pmatrix}. \tag{3}$$

As is mentioned above, we discuss the example of a massive scalar field theory. The propagator on the orbifold is represented as:

$$G_0^{(M)}(x, y) = \frac{1}{M} \sum_{m=0}^{M-1} \tilde{G}_0(x, y; m), \tag{4}$$

$$\tilde{G}_0(x, y; m) = \int \frac{d^{d+1}p}{(2\pi)^{d+1}} \frac{e^{i((\hat{g}^m p) \cdot x - p \cdot y) + i p_{\|} \cdot (x_{\|} - y_{\|})}}{p^2 + m^2}. \tag{5}$$

It is easy to see that $\tilde{G}_0$ with nonzero m does not conserve the perpendicular components of the momentum. Rather, it carries a *twisted* momentum: $(\boldsymbol{p}, p_{\parallel}) \to (\hat{g}^m \boldsymbol{p}, p_{\parallel})$. On the other hand, interaction vertices consist of coupling constants and the $(d+1)$-dimensional delta functions representing the momentum conservation. Note that the factor of the delta function for $\boldsymbol{p}$'s is invariant under a simultaneous rotation by $\hat{g}$:

$$\delta^2(\boldsymbol{p}_1 + \boldsymbol{p}_2 + \cdots) = \delta^2(\hat{g}(\boldsymbol{p}_1 + \boldsymbol{p}_2 + \cdots)). \qquad (6)$$

By using this invariance and Eqs. (4) and (5), we can prove that the general formula for a L-loop bubble diagram B, which contributes to the free energy, takes the following form:

$$B = \frac{V_{d-1}}{M} \sum_{m_1, \cdots, m_L = 0}^{M-1} \tilde{B}(m_1, \cdots, m_L), \qquad (7)$$

$$\tilde{B}(m_1, \cdots, m_L) = \int \prod_{l=1}^{L} [d^2 \boldsymbol{p}_l d^{d-1} p_{\parallel l}] I(p) \delta^2 \left(\sum_{l=1}^{L} (\hat{g}^{m_l} - 1) \boldsymbol{p}_l \right). \qquad (8)$$

Here, V_{d-1} denotes the area of the boundary ∂A, $V_{d-1} = \delta^{d-1}(0)$, whose divergence is not important in the following discussion. $\tilde{B}$ is a contribution from a fixed configuration of twists $(m_1 \cdots, m_l)$. $I(p)$ is an ordinary integrand for the diagram. It is essential that the configuration is characterized by the twists of the loop momenta.

Among the configurations of twists, the trivial one $(m_1 = 0, \cdots, m_L = 0)$ yields the delta function with vanishing argument. Therefore, the contribution to the free energy is proportional to $(V_{d-1} \times \delta^2(\boldsymbol{0})) = V_{d+1}$, the volume of the bulk. However, it does not contribute to EE, because its M-dependence comes only from the prefactor in Eq. (7), and it is canceled in Eq. (2). On the other hand, all the other configurations have nontrivial arguments in their delta functions, and they are proportional to the area of the boundary. By combining this fact with the dimensional analysis, we conclude that the leading contribution from B to EE is proportional to $V_{d-1}/\varepsilon^{d-1}$. Since this is a statement for the general bubble diagram, we have seen that the area-law for EE holds at all order.

3. Two specific contributions to the entanglement entropy

In principle, we can obtain the free energy on the orbifold by computing $\tilde{B}$'s and summing them over the twists configurations. This is, however, a technically difficult task even for diagrams with a few loops. Instead, we can

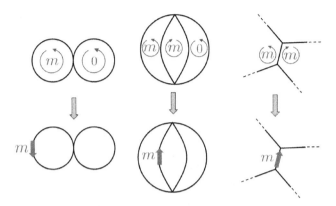

Fig. 2. Examples of the configurations of twists we call the propagator contribution. The blue cycles and letters denote the twists on the corresponding loop momenta. The twists in the upper diagrams are realized by the twists on a propagator as shown in the lower diagrams.

extract some specific contributions of physical importance. In the following, we discuss them and later their dominance in the whole contribution.

While the twists are acting on the loop momenta in each $\tilde{B}$, in some configurations we can attribute them to one twist acting on a momentum on a single propagator, as is shown in Fig. 2. Let us call such class of contributions *the propagator contribution*. The point is that we can sum them up in a systematic way, as is explained below, and can straightforwardly compute their contributions to EE. In those diagrams, most of the parts are identical to those in the flat space; only one propagator gets replaced with Eq. (5). As a naive attempt, let us sum them up by three steps: (i) considering the bubble diagrams in the flat space, (ii) taking the derivative of them with respect to the propagator, and (iii) reconnecting them with Eq. (5). Note that in Step (ii), we obtain the interaction correction part in the exact two-point function G. Then, their contributions to $F^{(M)}$ and EE would be summed up to take the following form:

$$F_{\text{prop}}^{(M)} \sim -\frac{1}{2} \int d^{d+1}x d^{d+1}y \left(G_0^{-1}(G - G)G_0^{-1} \right) \sum_{m=0}^{M-1} \tilde{G}_0(m), \qquad (9)$$

$$S_{\text{prop,int}} \sim \frac{V_{d-1}}{12} \int^{1/\varepsilon} \frac{d^{d-1}k_{\parallel}}{(2\pi)^{d-1}} \left[\Sigma G_0 + (\Sigma G_0)^2 + (\Sigma G_0)^3 + \cdots \right]. \quad (10)$$

In Eq. (9), we have omitted the arguments (x, y) in the both factors. In Eq. (10), we have represented the contribution in the momentum representation and expand it with respect to the 1PI part Σ. The arguments of all the terms in the square bracket are $(\mathbf{k} = \mathbf{0}, k_\parallel)$.

In fact, this is not correct and there is a subtlety in the attribution of the configurations to a twist on a single propagator. In some diagrams, the twists can be attributed more then one propagator. The method of taking the variation, as we have considered above, counts each of such attributions as distinct contributions. However, since what we have to sum up actually is the twist configuration, it leads to over-counting. Thus, Eq. (10) requires to be modified. Fortunately, we can show that the over-counting factors for diagrams are exactly classified according to the 1PI expansion. It means that we obtain the correct propagator contribution by dividing each term in Eq. (10) by the corresponding over-counting factor:

$$
S_{\text{prop,int}} = \frac{V_{d-1}}{12} \int^{1/\varepsilon} \frac{d^{d-1}k_\parallel}{(2\pi)^{d-1}} \left[\Sigma G_0 + \frac{1}{2}(\Sigma G_0)^2 + \frac{1}{3}(\Sigma G_0)^3 + \cdots \right]
$$

$$
= -\frac{V_{d-1}}{12} \int^{1/\varepsilon} \frac{d^{d-1}k_\parallel}{(2\pi)^{d-1}} \log(1 - \Sigma G_0). \tag{11}
$$

Furthermore, we can deal with the free part on the same foot. It is because its contribution to the free energy, $(1/2)\log G_0^{-1}$, is expanded by Schwinger parametrization to the series of ring-shaped diagrams, each of which consists of the propagators simply connected at their endpoints. As a result, their contribution to EE is identical to the one derived in the literatures. By combining it with Eq. (11), the propagator contribution is summed up to take the form of

$$
S_{\text{prop}} = \frac{V_{d-1}}{12} \int^{1/\varepsilon} \frac{d^{d-1}k_\parallel}{(2\pi)^{d-1}} \log G(\mathbf{k} = \mathbf{0}, k_\parallel). \tag{12}
$$

Note that the radiative corrections entering G have been calculated in the diagram in the flat space, and the final expression does not depend on the fact that we have introduced the orbifold in the middle of the analysis. After all, a potential danger of the additional counterterm on the singularity of the orbifold decouples, and we can consider the counterterms and renormalization just as in the flat space. Therefore, Eq. (12) is indeed the contribution written with the renormalized two-point function.

We also extract contributions to EE other than the propagator contribution. Consider a class of twist configurations as is shown in Fig. 3. In each of them, the twists are attributed to a twist of a loop momentum flowing at some channel in a single vertex. We call them *the vertex contribution*. It can be explicitly derived almost in the same way as the

131

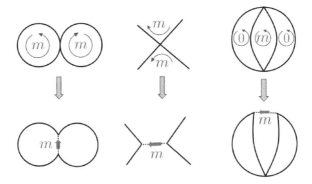

Fig. 3. Examples of the configurations of twists we call the vertex contribution. The twists in the upper diagrams are realized by the twists on a channel of a vertex as shown in the lower diagrams. The dot lines denotes the delta function to decompose the vertex.

propagator contribution.

We take the variation of bubble diagrams with respect to the vertex, namely the bare correlation function. It gives the exact correlation function. Here is a difference; we organize the endpoints into two, and reconnect them with the twisted version of the delta function. In general, there are choices of how to organize the endpoints. It corresponds to the channel we consider, at which the momentum gets twisted. Accordingly, the organization turns the correlation function into the two-point function of the composite operators. The connection by the twisted delta function comes from the fact that the momentum twist act on the channel of the vertex, which is expressed by decomposing the vertex with the delta function. In this case, we have the subtlety and over-counting, and have the corresponding 1PI-like expansion, and obtain the correct contribution by dividing the terms by the over-counting factor.

This computation results in the formula for the vertex contribution, which is written in terms of the renormalized two-point correlation function of the composite operators. For instance, in ϕ^4-theory, we have $:\phi^2:$ as the only composite operator, and the vertex contribution to EE is given by

$$S_{\text{vert}} = \frac{V_{d-1}}{12} \int^{1/\varepsilon} \frac{d^{d-1}k_{\parallel}}{(2\pi)^{d-1}} \log G_{\phi^2\phi^2}(\boldsymbol{k} = \boldsymbol{0}, k_{\parallel}), \tag{13}$$

where $G_{\phi^2\phi^2}$ is the two-point correlation function for $:\phi^2:$. In general interacting theories, we have several vertices and each of them has multiple

132

channel to be considered. These give rise to many composite operators, which are mixed with one another. Even in such cases, we can go through the same discussion and get a similar expression of the vertex contribution. The operator mixing is taken into account by representing the correlation functions in the form of a matrix, whose elements are correlation functions of operators. As for operators with nonzero spin, we have to take trace over the space of spins. It leads to the introduction of some coefficient matrix. Together with the propagator contribution, which is nothing but the contribution from the fundamental operators, the contribution we discuss is written as

$$S_{\text{prop+vert}} = \frac{V_{d-1}}{12} \int^{1/\varepsilon} \frac{d^{d-1}k_{\parallel}}{(2\pi)^{d-1}} \operatorname{tr} \hat{c} \log \hat{G}, \tag{14}$$

$$\hat{G}_{ab} = G_{\mathcal{O}_a \mathcal{O}_b}(\boldsymbol{k} = \boldsymbol{0}, k_{\parallel}). \tag{15}$$

Here, $G_{\mathcal{O}_a \mathcal{O}_b}$ denotes the renormalized correlation function of operators $\mathcal{O}_a$ and $\mathcal{O}_b$, that is normalized to be dimensionless by the cutoff ε. $\hat{c}$ is the numerical coefficient matrix, which is determined by traces over the spaces of spins. In the case where the operators are all scalar, $\hat{c}$ is a unit matrix.

The result shows that the contributions of EE we have discussed is understood in terms of the renormalized correlation functions of various operators. This is an all-order analysis, and implies that a part of divergence in naive calculation of EE comes from the ordinary radiative corrections, and in a sense irrelevant once we study EE in terms of renormalized physical quantities. On the other hand, the explicit appearance of the cutoff scale in Eq. (15) simply suggests that EE should be measured with some scale to be finite. Note that even in the free field theory, we have to introduce some cutoff scale to define EE. Therefore, it is rather a matter of the definition of EE.

There are many contributions to EE other than those we have discussed. They are expected, however, to be sub-leading, since all of the corresponding diagrams are of relatively higher loops. Their contributions to the free energy should be less dominant. In particular, when we consider a theory described by an Wilsonian effective action, where the quantum fluctuation has already been integrated out, such contributions are absent in the first place.[a] It is thus certain that Eq. (15) captures the dominant of EE.

---

[a]In that case, we have infinitely many operators from vertices in the effective actions. However, since the correlation between higher dimensional operators tends to have a shorter correlation length, we can approximate EE by a finite number of the low-dimensional operaotors.

4. Summary and outlook

In this talk, we have discussed EE in interacting field theories with the subregion of a half-space. We have reported in this case that a part of EE can be expressed in terms of renormalized two-point functions with the fundamental and composite operators. The composite operators arises in the decomposition of interaction vertices with all possible channels. This contribution to EE is certain to be dominant, and if we treat the Wilsonian effective action, it captures the whole EE. We expect that this result should be a significant step to associate EE to observations in a realistic physics.

The most crucial question is whether and how we can generalize the result to the cases of the general subregions. Of course, our analysis with the orbifold technique highly depends on the fact that the subregion is the flat half-space. On the other hand, the final result of Eq. (15) it self does not. We further expect that, the qualitative implication that the dominant part of EE is described with the two-point functions of operators, should hold in more general cases. The reason is that once we express Eq. (15) in the position space representation, we do not see the restriction of the form of the subregion. A possible way to investigate it is to reformulate things directly in the position space representation. It is a challenging but interesting future work. As another direction, It is important to study the structure of the renormalization in detail. While our analysis holds to all-order, it is within a perturbative picture. It is worthwhile to pursue a nonperturbative proof for our statement.

Acknowledgments

The author thanks to the organizers of East Asia Joint Symposium on Fields and Strings 2021 for giving the opportunity to talk. This work is partially supported by the Grant-in-Aid for JSPS Research Fellow, No. 20J00079.

References

1. S. Ryu and T. Takayanagi, Holographic derivation of entanglement entropy from AdS/CFT, *Phys. Rev. Lett.* **96**, p. 181602 (2006).
2. S. Ryu and T. Takayanagi, Aspects of Holographic Entanglement Entropy, *JHEP* **08**, p. 045 (2006).
3. V. E. Hubeny, M. Rangamani and T. Takayanagi, A Covariant holographic entanglement entropy proposal, *JHEP* **07**, p. 062 (2007).
4. V. Rosenhaus and M. Smolkin, Entanglement Entropy for Relevant and Geometric Perturbations, *JHEP* **02**, p. 015 (2015).

5. V. Rosenhaus and M. Smolkin, Entanglement Entropy: A Perturbative Calculation, *JHEP* **12**, p. 179 (2014).

6. V. Rosenhaus and M. Smolkin, Entanglement entropy, planar surfaces, and spectral functions, *JHEP* **09**, p. 119 (2014).

7. P. Calabrese and J. L. Cardy, Entanglement entropy and quantum field theory, *J. Stat. Mech.* **0406**, p. P06002 (2004).

8. H. Casini and M. Huerta, Entanglement entropy in free quantum field theory, *Journal of Physics A: Mathematical and Theoretical* **42**, p. 504007 (dec 2009).

9. A. Botero and B. Reznik, Spatial structures and localization of vacuum entanglement in the linear harmonic chain, *Phys. Rev. A* **70**, p. 052329 (Nov 2004).

10. D. Katsinis and G. Pastras, An Inverse Mass Expansion for Entanglement Entropy in Free Massive Scalar Field Theory, *Eur. Phys. J. C* **78**, p. 282 (2018).

11. E. Bianchi and A. Satz, Entropy of a subalgebra of observables and the geometric entanglement entropy, *Phys. Rev. D* **99**, p. 085001 (2019).

12. A. Lewkowycz, R. C. Myers and M. Smolkin, Observations on entanglement entropy in massive QFT's, *JHEP* **04**, p. 017 (2013).

13. C. P. Herzog and T. Nishioka, Entanglement Entropy of a Massive Fermion on a Torus, *JHEP* **03**, p. 077 (2013).

14. M. P. Hertzberg and F. Wilczek, Some Calculable Contributions to Entanglement Entropy, *Phys. Rev. Lett.* **106**, p. 050404 (2011).

15. S. Iso, T. Mori and K. Sakai, Entanglement Entropy in Interacting Field Theories (3 2021).

16. S. Iso, T. Mori and K. Sakai, Non-Gaussianity of Entanglement Entropy and Correlations of Composite Operators (5 2021).

17. S. Iso, T. Mori and K. Sakai, Wilsonian Effective Action and Entanglement Entropy, *Symmetry* **13**, p. 1221 (2021).

18. T. Nishioka and T. Takayanagi, AdS bubbles, entropy and closed string tachyons, *Journal of High Energy Physics* **2007**, 090 (jan 2007).

19. S. He, T. Numasawa, T. Takayanagi and K. Watanabe, Notes on Entanglement Entropy in String Theory, *JHEP* **05**, p. 106 (2015).

The AdS$_5 \times$ S^5 supercoset sigma model

from 4D Chern-Simons theory

Jun-ichi Sakamoto*

Department of Physics and Center for Theoretical Sciences,
National Taiwan University, Taipei 10617, Taiwan
* E-mail: sakamoto@ntu.edu.tw

In this paper, we explain how to derive the AdS$_5 \times$ S^5 supercoset sigma model in the Green-Schwarz formalism. The derivation is a generalization of the procedure for the PCM case developed by Delduc et al [arXiv:1909.13824]. This paper is based on the original paper [arXiv:2005.04950].

Keywords: Integrable systems; Sigma models; AdS$_5 \times$ S^5 supercoset sigma model.

1. Introduction

Integrable systems give a fascinating playground for investigating nonlinear dynamical systems. Determining whether a given theory is classical integrable requires to construct a Lax pair which ensures the existence of an infinite many conserved charges, but finding a Lax pair is a difficult task due to the lack of a guiding principle. Recently, a nice way to handle this issue has been proposed by Costello and Yamazaki[1] which is based on a 4D Chern-Simons (CS) theory[2,3] with a meromorphic 1-form ω. According to their proposal, by specifying the 1-form ω and the boundary conditions of the gauge field of the 4D CS theory, one can systematically construct a 2D classical integrable field theory. This Costello-Yamazaki proposal has been explored in subsequent works, and it has been shown that 4D CS theory can describe various 2D integrable field theories and integrable deformations including the Yang-Baxter (YB) deformation[4–9] and the λ-deformation[10,11]. For a recent review on this subject, see[12].

In this paper, we will explain how the AdS$_5 \times$ S^5 supercoset sigma model can be derived from a 4D CS theory by following the original paper[13] (see also[14]). As mentioned in the above, it is necessary to determine a memorphic 1-form ω. Remarkably, this 1-form ω is identified with a twist function characterizing the Poisson structure of the integrable system by Vicedo[15]. The Poisson structure of the $\mathrm{AdS}_5 \times \mathrm{S}^5$ supercoset sigma model has been studied in[16,17], and we can read off the twist function from their results. Then by considering a 4D CS theory with the twist function and specifying the boundary conditions we give the $\mathrm{AdS}_5 \times \mathrm{S}^5$ supercoset sigma model in the Green-Schwarz formalism. The derivation we use here is a generalization of the procedure developed by Delduc, Laxcroix, Magro and Vicedo[18].

This paper is organized as follows. Section 2 explains how to derive 2D NLSMs from 4D CS theory. In section 3, we derive the $\mathrm{AdS}_5 \times \mathrm{S}^5$ supercoset sigma model by generalizing the preceding result on the PCM. Section 4 is devoted to conclusion and discussion.

2. 2D integrable field theory from 4D CS theory

In this section, we will give a brief review about a derivation of 2D integrable field theories from a 4D CS theory by following[1,18].

2.1. 4D CS theory

We start with introducing the action of 4D CS theory. Let $G^{\mathbb{C}}$ be a complexified semisimple Lie group, and $\mathfrak{g}^{\mathbb{C}}$ be the Lie algebra for $G^{\mathbb{C}}$ with a non-degenerate symmetric bilinear form $\langle \cdot, \cdot \rangle : \mathfrak{g}^{\mathbb{C}} \times \mathfrak{g}^{\mathbb{C}} \to \mathbb{C}$. Then the action of the 4D CS theory is given by

$$S[A] = -\frac{i}{4\pi} \int_{\mathcal{M} \times \mathbb{C}P^1} \omega \wedge CS(A)\,, \tag{1}$$

where a 2D Minkowski space $\mathcal{M}$ has a metric $\eta_{ij} = \mathrm{diag}\,(-1, 1)$ with coordinates (τ, σ) and a global holomorphic coordinate of $\mathbb{C} \subset \mathbb{C}P^1 = \mathbb{C} \cup \{\infty\}$ is denoted by z. The gauge field A on $\mathcal{M} \times \mathbb{C}P^1$ is taking a value of $\mathfrak{g}^{\mathbb{C}}$, and $CS(A)$ is the CS 3-form

$$CS(A) \equiv \left\langle A, dA + \frac{2}{3} A \wedge A \right\rangle\,. \tag{2}$$

Here, ω is a meromorphic 1-form on $\mathbb{C}P^1$

$$\omega \equiv \varphi(z)dz\,, \tag{3}$$

where $\varphi(z)$ is a meromorphic function on $\mathbb{C}P^1$. It is noted that the meromorphic function $\varphi(z)$ can be identified with the twist function characterising the Poisson bracket associated with the 2D integrabe field theory[15]. For later use, we denote the sets of zeros and poles by $\mathfrak{z}$ and $\mathfrak{p}$, respectively.

The fact that ω is proportional to dz indicates that the action (1) is invariant under the transformation

$$A \mapsto A + \chi\, dz \,, \tag{4}$$

where χ is any $\mathfrak{g}^{\mathbb{C}}$-valued function defined on $\mathcal{M} \times \mathbb{C}P^1$. Hence, by using the gauge symmetry (4), the gauge field A can be expanded as

$$A = A_+ \, d\sigma^+ + A_- \, d\sigma^- + A_{\bar{z}} \, d\bar{z} \,. \tag{5}$$

where we introduced the light-cone coordinates defined as $\sigma^{\pm} \equiv \frac{1}{2}\left(\tau \pm \sigma\right)$. In the following discussion, we will ignore the $\bar{z}$-component of A.

Equations of motion

Next, let us give equations of motion for the action (1).

By taking a variation of the action (1) with respect to the gauge field, we obtain the bulk equations of motion

$$F_{+-} = 0 \,, \qquad \omega\, F_{\bar{z}\pm} = 0 \,, \qquad F(A) \equiv dA + A \wedge A \,, \tag{6}$$

and the boundary equation of motion

$$d\omega \wedge \langle A, \delta A \rangle = 0 \,. \tag{7}$$

Here, the factor ω in (6) is kept because there is a possibility that $\partial_{\bar{z}} A_{\pm}$ are distributions on $\mathbb{C}P^1$ supported by $\mathfrak{z}$. Note that the boundary equation of motion (7) has the support only on $\mathcal{M} \times \mathfrak{p} \subset \mathcal{M} \times \mathbb{C}P^1$. This follows from the fact that since $d\omega$ can be expanded as

$$d\omega = \partial_{\bar{z}}\varphi(z)\, d\bar{z} \wedge dz \,, \tag{8}$$

only the poles of φ can contribute to the distributions according to the relations

$$\delta(z - x) = \frac{1}{2\pi i} \frac{\partial}{\partial \bar{z}} \left(\frac{1}{z - x} \right) \,, \qquad x \in \mathfrak{p} \,. \tag{9}$$

2.2. Lax form

Classical integrablity of a given 2D classical field theory requires the existence of a Lax pair $\mathcal{L}$ on the 2D spacetime which is valued on $\mathfrak{g}^{\mathbb{C}}$, and satisfies the following properties: a) it is on-shell flat, b) it meromorphically depends on an auxiliary Riemann surface. In this subsection, we will explain how the Lax pair underlying a 2D integrable field theory is introduced in the context of the 4D CS theory.

By following[1,18], the Lax form $\mathcal{L}$ can be introduced by performing the formal gauge transformation

$$A = -d\hat{g}\hat{g}^{-1} + \hat{g}\,\mathcal{L}\,\hat{g}^{-1}\,, \qquad (10)$$

where a smooth function $\hat{g} : \mathcal{M} \times \mathbb{C}P^1 \to G^{\mathbb{C}}$, and we take a gauge choice

$$\mathcal{L}_{\bar{z}} = 0\,. \qquad (11)$$

In order to see whether the $\mathfrak{g}^{\mathbb{C}}$-valued 1-form $\mathcal{L}$ introduced in (10) satisfies properties a) and b), we rewrite the bulk equations of motion (6) in terms of $\mathcal{L}$ as

$$\partial_+\mathcal{L}_- - \partial_-\mathcal{L}_+ + [\mathcal{L}_+, \mathcal{L}_-] = 0\,, \qquad (12)$$

$$\omega \wedge \partial_{\bar{z}}\mathcal{L} = 0\,. \qquad (13)$$

The first constraint (12) is regarded as the on-shell flatness condition for the Lax form on $\mathcal{M}$. The second constraint (13) indicates that $\mathcal{L}$ is a meromorphic 1-form with poles at the zeros of ω. In this way, the 1-form $\mathcal{L}$ satisfies the properties a) and b), and can be regarded as the Lax form on $\mathcal{M}$ by identifying z on $\mathbb{C}P^1$ with a spectral parameter.

2.3. From 4D to 2D via the archipelago conditions

Here, let us give a general formula for the 2D action of the underlying integrable field theory by performing a dimensional reduction of the 4D action (1) along $\mathbb{C}P^1$.

For this purpose, by substituting (10) to (1), we rewrite the 4D action (1) in terms of $\mathcal{L}$ as

$$S[A] = -\frac{i}{4\pi}\int_{\mathcal{M}\times\mathbb{C}P^1}\omega \wedge I_{WZ}[\hat{g}] - \frac{i}{4\pi}\int_{\mathcal{M}\times\mathbb{C}P^1}\omega \wedge d\langle\hat{g}^{-1}d\hat{g}, \mathcal{L}\rangle\,. \qquad (14)$$

In order to reduce the 4D action (14) to a 2D one on $\mathcal{M}$, we need to impose the archipelago conditions[18] on $\hat{g}$. The archipelago conditions for $\hat{g}$ are defined as follows: There exist open disks V_x, U_x for each $x \in \mathfrak{p}$ such that $x \in V_x \subset U_x$ and

i) $U_x \cap U_y = \phi$ if $x \neq y$ for all $x, y \in \mathfrak{p}$,

ii) $\hat{g} = 1$ outside $M \times \cup_{x\in\mathfrak{p}}U_x$,

iii) $\hat{g}|_{\mathcal{M}\times U_x}$ depends only on $\sigma^\pm$ and the radial coordinate $|\xi_x|$ where ξ_x is the local holomorphic coordinate defined as $\xi_x \equiv z - x$,

iv) $\hat{g}|_{\mathcal{M}\times V_x}$ depends only on $\sigma^\pm$, that is, $\hat{g}_x \equiv \hat{g}_{\mathcal{M}\times V_x} = \hat{g}|_{\mathcal{M}\times\{x\}}$.

Here, we assume that $\hat{g}$ satisfies the archipelago condition (see, for example,[18] for the existence of such $\hat{g}$). Then, the 4D action can be reduced to the 2D action by performing an integral over $\mathbb{C}P^1$[18] as follows:

$$S\left[\{g_x\}_{x\in\mathfrak{p}}\right] = \frac{1}{2}\sum_{x\in\mathfrak{p}}\int_{\mathcal{M}}\left\langle \mathrm{res}_x(\varphi\,\mathcal{L}), g_x^{-1}dg_x\right\rangle$$

$$+\frac{1}{2}\sum_{x\in\mathfrak{p}}(\mathrm{res}_x\,\omega)\int_{\mathcal{M}\times[0,R_x]}I_{\mathrm{WZ}}\left[g_x\right]. \qquad (15)$$

The action (15) is invariant under the transformation

$$g_x \mapsto g_x h\,, \qquad \mathcal{L}\mapsto h^{-1}\mathcal{L}h + h^{-1}dh\,, \qquad (16)$$

where $h : \mathcal{M} \to G^{\mathbb{C}}$. The symmetry can be regarded as the residual gauge symmetry of the Lax form.

As discussed in[13], in order to determine a 2D integrable field theory, it is necessary to specify not only the boundary conditions, but also the behavior of $\hat{g}$. This point will be explained in the next section.

3. The GS action of the AdS$_5 \times$ S^5 supercoset sigma model from the 4D CS theory

In this section, we will give a derivation of the Green-Schwarz (GS) action of the AdS$_5 \times$ S^5 supercoset sigma model from the 4D CS theory by following[13].

3.1. The AdS$_5 \times$ S^5 supercoset sigma model

Here, let us review a supercoset construction of the AdS$_5 \times$ S^5 superstring.

The classical action of the AdS$_5 \times$ S^5 superstring in the GS formalism has been constructed based on the following supercoset[19]

$$\frac{PSU(2,2|4)}{SO(1,4)\times SO(5)}. \qquad (17)$$

As is well known, the super Lie algebra $\mathfrak{su}(2,2|4)$ can be decomposed to vector subspaces with respect to the $\mathbb{Z}_4$-grading structure:

$$\mathfrak{g} = \mathfrak{g}^{(0)} \oplus \mathfrak{g}^{(1)} \oplus \mathfrak{g}^{(2)} \oplus \mathfrak{g}^{(3)}\,, \qquad \mathfrak{g}^{(0)} = \mathfrak{so}(1,4)\times\mathfrak{so}(5)\,, \qquad (18)$$

where $\mathfrak{g}^{(0)} \oplus \mathfrak{g}^{(2)}$ and $\mathfrak{g}^{(1)} \oplus \mathfrak{g}^{(3)}$ are the bosonic and fermionic parts of $\mathfrak{su}(2,2|4)$, respectively, and the commutation relations of $\mathfrak{g}^{(m)}$ satisfy

$$[\mathfrak{g}^{(m)},\mathfrak{g}^{(n)}] \subset \mathfrak{g}^{(k)} \qquad (m+n=k \mod 4)\,. \qquad (19)$$

140

Note that the $\mathbb{Z}_4$-graded property ensures the classical integrablity of the $\text{AdS}_5 \times \text{S}^5$ superstring[20].

The GS action (32) of the $\text{AdS}_5 \times \text{S}^5$ supercoset sigma model is given by

$$S[g] = \int_{\mathcal{M}} \text{Str} \left(j_- d_+(j_+) \right) d\sigma^+ \wedge d\sigma^- , \tag{20}$$

where $j = g^{-1}dg \in \mathfrak{su}(2,2|4)$ is the left-invariant current for $g \in SU(2,2|4)$. Here, $d_\pm$ are the linear combinations of the projection operators $P_{(k)} : \mathfrak{g} \to \mathfrak{g}^{(k)}$ ($k = 0,1,2,3$) like

$$d_\pm = \pm P_{(1)} + 2P_{(2)} \mp P_{(3)} . \tag{21}$$

The associated Lax pair is[20]

$$\mathcal{L} = \left(z^{-1} j_+^{(3)} + j_+^{(0)} + z\, j_+^{(1)} + z^2\, j_+^{(2)} \right) d\sigma^+$$
$$+ \left(z^{-2}\, j_-^{(2)} + z^{-1}\, j_-^{(3)} + j_-^{(0)} + z\, j_-^{(1)} \right) d\sigma^- . \tag{22}$$

We can see that the on-shell flatness condition for (22) is equivalent to the equations of motion of the action (20).

The Poisson structure of the $\text{AdS}_5 \times \text{S}^5$ superstring has been considered in[16,17]. By computing the Poisson brackets of the Lax pair (22), we can obtain the twist function of the $\text{AdS}_5 \times \text{S}^5$ supercoset sigma model[a]

$$\varphi_{\text{str}}(z) = \frac{4z^3}{(z^4 - 1)^2} . \tag{23}$$

The twist function (23) has the following poles and zeros:

$$\mathfrak{p} = \{+1, -1, +i, -i\} , \qquad \mathfrak{z} = \{0, \infty\} , \tag{24}$$

where the poles are double poles and the zeros are triple zeros, respectively.

3.2. A derivation from the 4D CS theory

Now, let us reproduce the GS action (20) of the $\text{AdS}_5 \times \text{S}^5$ supercoset sigma model from the 4D CS theory (14) with the meromorphic 1-form

$$\omega = \varphi_{\text{str}}(z)\, dz . \tag{25}$$

---

[a] $\varphi_{\text{str}}(z)$ is slightly different from $\phi_{\text{string}}(z)$ in (2.10) of[8]. These are related via $\varphi_{\text{str}}(z) = \frac{1}{z} \phi_{\text{string}}(z)$.

In this case, the gauge field A in (14) takes a value in $\mathfrak{g} = \mathfrak{su}(2,2|4)$, and the bracket $\langle \cdot, \cdot \rangle$ in the 4D action (1) is replaced by the supertrace Str.

As mentioned in the previous section, we need to choose a solution to the boundary equation of motion to specify a 2D integrable model. The associated boundary equations of motion for the meromorphic 1-form (25) are

$$\epsilon^{ij} \langle\!\langle (A_i, \partial_{\xi_p} A_i), \delta(A_j, \partial_{\xi_p} A_j) \rangle\!\rangle_p = 0, \qquad p \in \mathfrak{p}, \tag{26}$$

where the double bracket is defined as

$$\langle\!\langle (x,y),(x',y') \rangle\!\rangle_p \equiv (\mathrm{res}_p \, \omega) \, \mathrm{Str}(x \cdot x') + (\mathrm{res}_p \, \xi_p \omega) \, (\mathrm{Str}(x \cdot y') + \mathrm{Str}(x' \cdot y))$$
$$= \frac{p}{4} \left(\mathrm{Str}(x \cdot y') + \mathrm{Str}(x' \cdot y) \right). \tag{27}$$

A possible solution to the boundary equationa of motion (26) is given by

$$A|_{z=p} = 0 \qquad (p \in \mathfrak{p}). \tag{28}$$

This boundary condition obviously solves the boundary equation of motion (26) and leads to the GS action of the $\mathrm{AdS}_5 \times \mathrm{S}^5$ supercoset sigma model as we will see in the following discussion.

Next, let us construct the associated Lax pair by solving the constraint (13) with the boundary condition (28). For our purpose, we consider the following ansatz for the Lax pair as

$$\mathcal{L} = \left(z^{-1} V_+^{[-1]} + V_+^{[0]} + z \, V_+^{[1]} + z^2 \, V_+^{[2]} \right) d\sigma^+$$
$$+ \left(z^{-2} V_-^{[-2]} + z^{-1} V_-^{[-1]} + V_-^{[0]} + z \, V_-^{[1]} \right) d\sigma^-, \tag{29}$$

where $V_{\pm}^{[n]}$ $(n = -1, 0, 1), V_{\pm}^{[\pm 2]} : \mathcal{M} \to \mathfrak{su}(2,2|4)$ are undetermined functions. It is easy to see that the ansatz (29) satisfies the bulk equation of motion (6) with the twist function (23). By solving the relation (10) between the Lax pair and the gauge field at each pole under boundary conditions (28), $V_{\pm}^{[k]}$ are determined as follows

$$V_{\pm}^{[0]} = \frac{j_{1,\pm} + j_{2,\pm} + j_{3,\pm} + j_{4,\pm}}{4}, \qquad V_{\pm}^{[\pm 2]} = \frac{j_{1,\pm} - j_{2,\pm} + j_{3,\pm} - j_{4,\pm}}{4},$$
$$V_{\pm}^{[1]} = \frac{j_{1,\pm} - i \, j_{2,\pm} - j_{3,\pm} + i \, j_{4,\pm}}{4}, \qquad V_{\pm}^{[-1]} = \frac{j_{1,\pm} + i \, j_{2,\pm} - j_{3,\pm} - i \, j_{4,\pm}}{4},$$
$$\tag{30}$$

where we denoted the left-invariant curret $\hat{g}^{-1}d\hat{g}$ at each pole of the twist function (23) by

$$
\begin{aligned}
j_1 &= \hat{g}^{-1}d\hat{g}|_{z=1}, \quad j_2 = \hat{g}^{-1}d\hat{g}|_{z=i}, \\
j_3 &= \hat{g}^{-1}d\hat{g}|_{z=-i}, \quad j_4 = \hat{g}^{-1}d\hat{g}|_{z=-1}.
\end{aligned}
\tag{31}
$$

Note that the above ansatz (29) is not the only possible solution that satisfies the boundary conditions (28). In fact, we can also consider Lax pairs for the pure spinor form[1].

Next, we substitute the Lax pair (29) into the master formula (15), and then obtain the 2D action

$$
\begin{aligned}
S[g_k] = \frac{1}{16}\int_{\mathcal{M}} \mathrm{Str}\Big[&\sum_{\sigma\in S^4} \Big(j_{\sigma(1),+} - (1+i)j_{\sigma(2),+} + j_{\sigma(3),+} - (1-i)j_{\sigma(4),+} \Big)j_{\sigma(1),-} \\
&- \Big(-j_{\sigma(1),-} + (1-i)j_{\sigma(2),-} - j_{\sigma(3),-} + (1+i)j_{\sigma(4),-} \Big)j_{\sigma(1),+} \Big]d\sigma^+ \wedge d\sigma^-,
\end{aligned}
\tag{32}
$$

where $\sigma \in S^4$ is a cyclic permutation of the set $\{1,2,3,4\}$. The action (32) is clearly invariant under the cyclic permutations of j_k.

As discussed in[13], it is necessary to impose a relation between j_k ($k = 1, \ldots, 4$), so that the 2D action (32) can reproduce the GS action of the $\mathrm{AdS}_5 \times \mathrm{S}^5$ supercoset sigma model. This can be achieved by requiring the following relation, which respects the cyclic symmetry of the 2D action (32):

$$
j_k = f_s^{k-1}(j) \qquad (k = 1, \ldots, 4),
\tag{33}
$$

where $j \in \mathfrak{su}(2,2|4)$ is the left-invariant current for $g \in SU(2,2|4)$, and the map $f_s : \mathfrak{su}(2,2|4)^{\mathbb{C}} \to \mathfrak{su}(2,2|4)^{\mathbb{C}}$ is an automorphism of $\mathfrak{su}(2,2|4)$ satisfying the $\mathbb{Z}_4$-grading property $f_s^4 = \mathrm{Id}$. For our purpose, we will take the $\mathbb{Z}_4$-grading automorphism f_s such that each subspace $\mathfrak{g}^{(k)}$ ($k = 0,1,2,3$) is the eigenspace of f_s satisfying

$$
f_s(\mathfrak{g}^{(k)}) = i^k \mathfrak{g}^{(k)}.
\tag{34}
$$

The explicit expression of f_s can be written down after taking a supermatrix realization of $\mathfrak{su}(2,2|4)$ (For the details, see[21]).

By imposing the constraints (33) satisfying (34), we can easily show that the 2D action (32) and the Lax pair (29) reduce to the ones (20), (22) of the $\mathrm{AdS}_5 \times \mathrm{S}^5$ supercoset sigma model in the GS formalism, respectively. In this way, the 4D CS theory with the moromorphic 1-form (25) can describe the $\mathrm{AdS}_5 \times \mathrm{S}^5$ supercoset sigma model.

4. Conclusion and Discussion

In this paper, we have reviewed how a 2D classical integrable field theory can be systematically derived from a 4D CS theory by following [18]. Then, by generalizing the previous procedure on the PCM case, we have reproduced the $AdS_5 \times S^5$ supercoset sigma model from a 4D CS theory with a meromorphic one-form (25). In addition, we should note a recent work [14] which has succeeded in introducing the world-sheet metric into an underlying 2D integrable field theory by extending a 4D CS theory. Thus, we can treat $AdS_5 \times S^5$ superstring theory beyond the $AdS_5 \times S^5$ supercoset sigma model in the context of the 4D CS theory.

It has been investigated how 4D CS theory describe various 2D classical integrable theories. As a next step, it is natural to examine ways to quantize 2D classical integrable field theories and their quantum integrability in this framework. The problem of quantization has been extensively investigated for so-called ultralocal models, such as the Faddeev-Reshetikhin model [22] and massless Thirring model. Note that these 2D theories can be derived from 4D CS theories with a meromorphic 1-form ω without zeroes [1,24,25]. It is known that ultralocal integrable field theories can be described by taking the appropriate continuous limit of integrable lattice models [22,23]. Since 4D CS theory can also describe integrable lattice models in terms of the expectation value of the Wilson loops [2,3], it would be an interesting problem to understand this picture in the framework of 4D CS theory.

When ω has zeros, e.g. PCM or coset sigma model, the quantization of the corresponding classical integrable theory becomes more subtle (For this issue, see [2,15] in the context of the 4D CS theory). Quantizing non-ultralocal classical integrable theories in a first-principles way is still an open problem and an important task.

Acknowledgments

We are grateful to Osamu Fukushima and Kentaroh Yoshida for the collaboration of the original work. The work of J.S. was supported in part by Ministry of Science and Technology (project no. 109-2811-M-002-539), National Taiwan University.

References

1. K. Costello and M. Yamazaki, "Gauge Theory And Integrability, III," arXiv:1908.02289 [hep-th].

2. K. Costello, E. Witten and M. Yamazaki, "Gauge Theory and Integrability, I," ICCM Not. 6, 46-191 (2018) [arXiv:1709.09993 [hep-th]].

3. K. Costello, E. Witten and M. Yamazaki, "Gauge Theory and Integrability, II," ICCM Not. 6, 120-149 (2018) [arXiv:1802.01579 [hep-th]].

4. C. Klimcik, "Yang-Baxter sigma models and dS/AdS T duality," JHEP **0212** (2002) 051 [hep-th/0210095].

5. C. Klimcik, "On integrability of the Yang-Baxter sigma-model," J. Math. Phys. **50** (2009) 043508 [arXiv:0802.3518 [hep-th]].

6. F. Delduc, M. Magro and B. Vicedo, "On classical q-deformations of integrable sigma-models," JHEP **1311** (2013) 192 [arXiv:1308.3581 [hep-th]].

7. F. Delduc, M. Magro and B. Vicedo, "An integrable deformation of the AdS_5xS^5 superstring action," Phys. Rev. Lett. **112** (2014) no.5, 051601 [arXiv:1309.5850 [hep-th]].

8. F. Delduc, M. Magro and B. Vicedo, "Derivation of the action and symmetries of the q-deformed $AdS_5 \times S^5$ superstring," JHEP **1410** (2014) 132 [arXiv:1406.6286 [hep-th]].

9. I. Kawaguchi, T. Matsumoto and K. Yoshida, "Jordanian deformations of the AdS_5xS^5 superstring," JHEP **1404** (2014) 153 [arXiv:1401.4855 [hep-th]].

10. K. Sfetsos, "Integrable interpolations: From exact CFTs to non-Abelian T-duals," Nucl. Phys. B **880** (2014) 225 [arXiv:1312.4560 [hep-th]].

11. T. J. Hollowood, J. L. Miramontes and D. M. Schmidtt, "Integrable Deformations of Strings on Symmetric Spaces," JHEP **1411** (2014) 009 [arXiv:1407.2840 [hep-th]].

12. S. Lacroix, "4-dimensional Chern-Simons theory and integrable field theories," [arXiv:2109.14278 [hep-th]].

13. O. Fukushima, J. Sakamoto and K. Yoshida, "Yang-Baxter deformations of the $AdS_5 \times S^5$ supercoset sigma model from 4D Chern-Simons theory," JHEP **09** (2020), 100 [arXiv:2005.04950 [hep-th]].

14. K. Costello and B. Stefański, "Chern-Simons Origin of Superstring Integrability," Phys. Rev. Lett. **125** (2020) no.12, 121602 [arXiv:2005.03064 [hep-th]].

15. B. Vicedo, "Holomorphic Chern-Simons theory and affine Gaudin models," arXiv:1908.07511 [hep-th].

16. H. Itoyama and T. Oota, "The AdS(5) x S**5 superstrings in the generalized light-cone gauge," Prog. Theor. Phys. **117** (2007), 957-972 [arXiv:hep-th/0610325 [hep-th]].

17. B. Vicedo, "Hamiltonian dynamics and the hidden symmetries of the $AdS_5 \times S^5$ superstring," JHEP **1001** (2010) 102 [arXiv:0910.0221 [hep-th]].

18. F. Delduc, S. Lacroix, M. Magro and B. Vicedo, "A unifying 2d action for integrable σ-models from 4d Chern-Simons theory," arXiv:1909.13824 [hep-th].

19. R. R. Metsaev and A. A. Tseytlin, "Type IIB superstring action in AdS(5) x S**5 background," Nucl. Phys. B **533** (1998) 109 [hep-th/9805028].

20. I. Bena, J. Polchinski and R. Roiban, "Hidden symmetries of the AdS(5) x S**5 superstring," Phys. Rev. D **69** (2004) 046002 [hep-th/0305116].

21. G. Arutyunov and S. Frolov, "Foundations of the $AdS_5 \times S^5$ Superstring. Part I," J. Phys. A **42** (2009) 254003 [arXiv:0901.4937 [hep-th]].

22. L. D. Faddeev and N. Y. Reshetikhin, "Integrability of the Principal Chiral Field Model in (1+1)-dimension," Annals Phys. **167** (1986), 227

23. C. Destri and H. J. de Vega, "Light Cone Lattices and the Exact Solution of Chiral Fermion and σ Models," J. Phys. A **22** (1989), 1329

24. O. Fukushima, J. Sakamoto and K. Yoshida, "Faddeev-Reshetikhin model from a 4D Chern-Simons theory," JHEP **02** (2021), 115 [arXiv:2012.07370 [hep-th]].

25. V. Caudrelier, M. Stoppato and B. Vicedo, "On the Zakharov-Mikhailov action: 4d Chern-Simons origin and covariant Poisson algebra of the Lax connection," Lett. Math. Phys. **111** (2021), 82 [arXiv:2012.04431 [hep-th]].

Target Space Entanglement
in Quantum Mechanics of Fermions and Matrices

Sotaro Sugishita*

Institute for Advanced Research, Nagoya University
and
Department of Physics, Nagoya University,
Nagoya, Aichi 464-8601, Japan
* E-mail: sugishita.sotaro.r6@f.mail.nagoya-u.ac.jp

Quantum entanglement is closely related to the structure of spacetime in quantum gravity. For quantum field theories or statistical models, we usually consider base space entanglement. However, target space instead of base space sometimes directly connects to our spacetime. In these cases, it is natural to consider a concept of target space entanglement. To define the target space entanglement, we consider a generalized definition of entanglement entropy based on an algebraic approach. This approach is reviewed and is applied to the first quantized particles, in particular, fermions. This article is based on the paper JHEP 08 (2021) 046[1].

Keywords: Entanglement; Matrix models; Target space; Mutual information.

1. Introduction

It is widely believed that quantum entanglement is closely related to the structure of spacetime in quantum gravity. In the AdS/CFT correspondence, the Ryu-Takayanagi formula[2] states that entanglement about the base space in holographic CFTs is connected to the area of minimal surface in the bulk. As in this example, we often consider the base space entanglement in quantum field theories or statistical models. However, target space instead of base space sometimes directly connects to our spacetime, for example, perturbative string theories or matrix models. Thus, it is natural to investigate a notion of target space entanglement[3-5]. See also recent Refs. 1,6-9.[a]

In Ref. 3, the target space entanglement is defined using an algebraic approach. We will review this approach in Sec. 2, and apply it to quantum mechanics of fermions in Sec. 3, Sec. 4 and Sec. 5.

---

[a]A concept of entanglement in string theories (matrix models) is investigated in 10 and revisited in 11.

148

2. Definition of entanglement entropy based on subalgebras of operators

Let us recall the conventional definition of entanglement entropy (EE). Suppose that a total density matrix ρ is given for a Hilbert space $\mathcal{H} = \mathcal{H}_B \otimes \mathcal{H}_{\bar{B}}$. The EE for subsystem $\mathcal{H}_B$ is defined as the von Neumann entropy of the reduced density matrix $\rho_B = \operatorname{tr}_{\bar{B}} \rho$ as $S_B = -\operatorname{tr}_B \rho_B \log \rho_B$. This definition relies on the tensor product structure of the Hilbert space, $\mathcal{H} = \mathcal{H}_B \otimes \mathcal{H}_{\bar{B}}$. However, total Hilbert spaces sometimes do not have such simple tensor-factorized forms. For example, the Hilbert space of a first-quantized (non-relativistic) particle in a space $\mathbb{R}^d$ is schematically given by a "direct sum" as $\mathcal{H} = \operatorname{span}\{|x\rangle|\, x \in \mathbb{R}^d\}$. Thus, even if we divide the space $\mathbb{R}^d$ into two subregions $\mathbb{R}^d = B \cup \bar{B}$, it is difficult to take the "partial trace" on $\bar{B}$.

The algebraic approach enables us to define EE without relying on the tensor product structure (see, e.g., the references [1,12–14]). The algebraic definition is based on the subalgebra of operators (observables). If a total density matrix ρ is given, and we have a restricted set of operators (subalgebra $\mathcal{A}$), an entropy $S_{\mathcal{A}}(\rho)$ associated with the subalgebra $\mathcal{A}$ is defined. This concept is natural, if we recall the meaning of entropy in information theory. The entropy is a measure of uncertainty about the whole information when we can only know partial information. If an observer can use only a subset of operators $\mathcal{A}$, the whole information is not obtained. Entropy $S_{\mathcal{A}}(\rho)$ quantifies the amount of uncertainty (or unknownness). In this sense, the usual EE, $S_B = -\operatorname{tr}_B \rho_B \log \rho_B$, for $\mathcal{H} = \mathcal{H}_B \otimes \mathcal{H}_{\bar{B}}$ represents uncertainty for an observer who can probe only subsystem $\mathcal{H}_B$. That is, it is the entropy for the subalgebra $\mathcal{L}(\mathcal{H}_B) \otimes 1_{\mathcal{H}_{\bar{B}}}$.[b] The choice of subalgebra $\mathcal{A}$ is arbitrary, and we do not need the tensor product structure.

For general subalgebra $\mathcal{A}$, the entropy $S_{\mathcal{A}}(\rho)$ is computed as follows. First, the 'reduced density matrix' $\rho_{\mathcal{A}}$ is uniquely determined from ρ and $\mathcal{A}$ as an operator in $\mathcal{A}$ satisfying the following equation:

$$\operatorname{tr}(\rho_{\mathcal{A}}\mathcal{O}) = \operatorname{tr}(\rho_{\mathcal{A}}\rho), \qquad {}^{\forall}\mathcal{O} \in \mathcal{A}. \tag{1}$$

For example, if the total Hilbert space has a tensor product form as $\mathcal{H} = \mathcal{H}_B \otimes \mathcal{H}_{\bar{B}}$, and we take the subalgebra $\mathcal{A}$ as $\mathcal{A} = \mathcal{L}(\mathcal{H}_B) \otimes 1_{\mathcal{H}_{\bar{B}}}$, then $\rho_{\mathcal{A}}$ is given by $\rho_B \otimes 1_{\mathcal{H}_{\bar{B}}}/\dim \mathcal{H}_{\bar{B}}$. The point is that the definition Eq. (1) is

---

[b]Here, $\mathcal{L}(V)$ denotes a set of linear operators on linear space V, and 1_V does the identity operator on V.

applicable even when the Hilbert space does not have the tensor product structure.

Furthermore, for a given subalgebra, we can decompose the Hilbert space into blocks of tensor products where the subalgebra acts nontrivially only on each tensor component as follows:

$$\mathcal{H} = \bigoplus_k \mathcal{H}_{B_k} \otimes \mathcal{H}_{\bar{B}_k} \quad \text{s.t.} \quad \mathcal{A} = \bigoplus_k \mathcal{L}(\mathcal{H}_{B_k}) \otimes 1_{\bar{B}_k}. \tag{2}$$

This decomposition is uniquely fixed by the subalgebra $\mathcal{A}$. We represents the projection onto each block by Π_k. We define the density matrix ρ_k on the projected space $\Pi_k \mathcal{H}$ as

$$\rho_k := \frac{1}{p_k} \Pi_k \rho \Pi_k, \tag{3}$$

where p_k is a normalization factor defined as $p_k := \text{tr}(\Pi_k \rho \Pi_k)$ and is a probability of being in the sector $\Pi_k \mathcal{H}$ for the given ρ. Since the projected space $\Pi_k \mathcal{H}$ has a simple tensor-factorized form as $\Pi_k \mathcal{H} = \mathcal{H}_{B_k} \otimes \mathcal{H}_{\bar{B}_k}$ in the decomposition (2), we can consider the reduced density matrix of ρ_k on $\mathcal{H}_{B_k}$ as

$$\rho_{B_k} := \text{tr}_{\bar{B}_k} \rho_k. \tag{4}$$

Then, the 'reduced density matrix' $\rho_{\mathcal{A}}$ satisfying Eq. (1) is given by

$$\rho_{\mathcal{A}} = \bigoplus_k p_k \, \rho_{B_k} \otimes \frac{1_{\bar{B}_k}}{\dim(\mathcal{H}_{\bar{B}_k})}. \tag{5}$$

We define the reduced density matrix ρ_B on space $\mathcal{H}_B = \bigoplus_k \mathcal{H}_{B_k}$ as

$$\rho_B := \bigoplus_k p_k \rho_{B_k}. \tag{6}$$

EE $S_{\mathcal{A}}(\rho)$ is defined as the von Neumann entropy

$$S_{\mathcal{A}}(\rho) = - \text{tr}_B \, \rho_B \log \rho_B = - \sum_k p_k \log p_k + \sum_k p_k S(\rho_{B_k}), \tag{7}$$

where $S(\rho_{B_k}) := - \text{tr}_{B_k} \rho_{B_k} \log \rho_{B_k}$. The first term in the r.h.s. of Eq. (7) is called the classical part,

$$S_{cl}(\rho, \mathcal{A}) := - \sum_k p_k \log p_k, \tag{8}$$

and is the Shannon entropy of the probability distribution $\{p_k\}$. On the other hand, the second term in the r.h.s. of Eq. (7) is called the quantum part $S_q(\rho, \mathcal{A})$. The expression in Eq. (7) is similar to the symmetry resolved entanglement entropy[15,16].

2.1. *Example: Entanglement in a single qubit*

As a concrete example of EE in the algebraic approach, we consider a single qubit. The Hilbert space is two-dimensional space, $\mathcal{H} = \mathrm{span}\{|0\rangle, |1\rangle\}$. We usually consider entanglement between two qubits. The algebraic approach enables us to consider "EE" even for a single qubit.

The full set of operators $\mathcal{L}(\mathcal{H})$ is $\mathcal{L}(\mathcal{H}) = \mathrm{span}\{I, \sigma_x, \sigma_y, \sigma_z\}$.[c] If we take the subalgebra as this full algebra, the decomposition Eq. (2) is trivial as

$$\mathcal{H} = \mathcal{H} \otimes \mathbb{C} \tag{9}$$

with $\mathcal{A} = \mathcal{L}(\mathcal{H}) \otimes 1$. In this case, ρ_B in Eq. (6) is just the original ρ. Thus, the EE associated with the full algebra $\mathcal{L}(\mathcal{H})$ is just the von Neumann entropy of ρ,

$$S_{\mathcal{L}(\mathcal{H})}(\rho) = -\operatorname{tr} \rho \log \rho. \tag{10}$$

In particular, if state ρ is pure, the entropy vanishes as $S_{\mathcal{L}(\mathcal{H})}(\rho) = 0$. It means that the pure state is not ambiguous and is completely determined by quantum tomography if we can use any operators.

Situation changes when we can use only a subset of operators. Let us suppose that we can probe only z-direction. This corresponds to taking subalgebra $\mathcal{A} = \mathrm{span}\{1, \sigma_z\}$. The decomposition (2) for this choice of the subalgebra is $\mathcal{H} = \mathrm{span}\{|0\rangle\} \oplus \mathrm{span}\{|1\rangle\}$ where $\mathcal{A} = \mathrm{span}\{1, \sigma_z\}$ can be represented as $\mathcal{A} = \mathrm{span}\left\{\begin{pmatrix} 1 & 0 \\ 0 & 0 \end{pmatrix}\right\} \oplus \mathrm{span}\left\{\begin{pmatrix} 0 & 0 \\ 0 & 1 \end{pmatrix}\right\}$. The projection Π_k are $\Pi_k = |k\rangle\langle k|$ $(k = 0, 1)$. We then have $p_k = \langle k|\rho|k\rangle$ and $\rho_{B_k} = \Pi_k$. Thus, the EE associated with the subalgebra $\mathcal{A}$ is

$$S_{\mathcal{A}}(\rho) = -p_0 \log p_0 - p_1 \log p_1, \tag{11}$$

where the quantum part $S_q(\rho, \mathcal{A})$ always vanishes, and the entropy is just the classical Shannon entropy of the probability distribution that the qubit is measured in 0 or 1 for the given state ρ. Even pure states in general have non-vanishing entropy (except for the case where states are eigenstates of σ_z). The non-vanishing entropy reflects the fact that pure states are ambiguous for restricted observers who can probe only z-direction. In fact, the observers cannot distinguish pure states with mixed states $\rho = \begin{pmatrix} p_0 & 0 \\ 0 & p_1 \end{pmatrix}$.

---

[c]We take a basis such that $\sigma_z = \begin{pmatrix} 1 & 0 \\ 0 & -1 \end{pmatrix}$ with $\sigma_z|k\rangle = (-1)^k|k\rangle$ $(k = 0, 1)$.

3. Entanglement of fermions with a fixed number

We now consider target space entanglement of first-quantized N fermions by the algebraic approach. The Hilbert space of the single particle is represented by $\mathcal{H}^{(1)}$. It is given by $\mathcal{H}^{(1)} = \text{span}\{|x\rangle | x \in M\}$ where M is the target space of particles. The Hilbert space $\mathcal{H}^{(N)}$ of N fermions is given by the N-th exterior power of $\mathcal{H}^{(1)}$ as

$$\mathcal{H}^{(N)} = \bigwedge^N \mathcal{H}^{(1)}, \tag{12}$$

which is spanned as $\mathcal{H}^{(N)} = \text{span}\{|x_1\rangle \wedge \cdots \wedge |x_N\rangle | x_i \in M (i = 1, \ldots, N)\}$.

We take a subregion B in the target space M, and consider the EE of this subregion. Since the Hilbert space $\mathcal{H}^{(N)}$ does not have a tensor-factorized structure with respect to the target space coordinates, we adopt the algebraic approach instead of the conventional definition. The subalgebra we take is the set of operators acting non-trivially only on particles in subregion B, which is represented by $\mathcal{A}(B)$. For example, when $N = 1$, the subalgebra $\mathcal{A}(B)$ is given by $\mathcal{A}(B) = \text{span}\{|y\rangle\langle y'| | y, y' \in B\} \oplus \text{span}\{\int_{\bar{B}} dz |z\rangle\langle z|\}$ where $\bar{B}$ is the complement region of B. For general N, we can decompose $\mathcal{H}^{(N)}$ into a direct sum of the following subsectors as

$$\mathcal{H}^{(N)} = \bigoplus_{k=0}^{N} \mathcal{H}_k^{(N)}. \tag{13}$$

The subsector $\mathcal{H}_k^{(N)}$ consists of states where k particles in B and $N - k$ ones in $\bar{B}$ as

$$\mathcal{H}_k^{(N)} = \text{span}\left\{|x_1\rangle \wedge \cdots \wedge |x_N\rangle \,|\, x_1, \ldots, x_k \in B, \, x_{k+1}, \ldots, x_N \in \bar{B}\right\}. \tag{14}$$

To represent the subalgebra $\mathcal{A}(B)$, we introduce the following abbreviated notation:

$$|\{x\}_n\rangle = |x_1\rangle \wedge \cdots \wedge |x_n\rangle \in \bigwedge^n \mathcal{H}^{(1)}. \tag{15}$$

The subalgebra $\mathcal{A}(B)$ is then given by

$$\mathcal{A}(B) = \bigoplus_{k=0}^{N} \mathcal{A}_k, \tag{16}$$

where $\mathcal{A}_k$ is a subalgebra on $\mathcal{H}_k^{(N)}$ and takes the form

$$\mathcal{A}_k = \text{span}\left\{\int_{\bar{B}} dz_1 \ldots dz_{N-k}(|\{y\}_k\rangle \wedge |\{z\}_{N-k}\rangle)(\langle\{y'\}_k| \wedge \langle\{z\}_{N-k}|)\right\}$$

with $y_1, \ldots, y_k, y_1', \ldots, y_k' \in B.$ \tag{17}

Since the subalgebra $\mathcal{A}(B)$ is specified, we can compute the entropy $S_{\mathcal{A}(B)}$ associated with this subalgebra in the manner described in the previous section.[d] We call this entropy the target space entanglement entropy S_B because the subalgebra $\mathcal{A}(B)$ is characterized by the subregion B in the target space of particles. In the second quantized picture, we can define the conventional entanglement entropy for subregion B. The target space EE S_B agrees with this base space EE[1,3,4].

4. Fermions in the Slater determinant states

To be more specific, we focus on pure states whose N-body wave functions are given by the Slater determinants as

$$\psi(x_1, \ldots, x_N) = \frac{1}{\sqrt{N!}} \det[\chi_i(x_j)] \qquad (i, j = 1, \ldots, N), \tag{18}$$

where $\chi_i(x)$ are one-body wave functions normalized as

$$\int_M dx \, \chi_i^*(x) \chi_j(x) = \delta_{ij}. \tag{19}$$

The target space EE for subregion B can be evaluated as Eq. (7) by computing p_k and $S(\rho_{B_k})$ for the pure states ψ. After some computations (see Ref. 1 for details), we can find that the entropy S_B follows the simple formula:

$$S_B(\psi) = -\operatorname{tr}[X \log X + (1_N - X) \log(1_N - X)], \tag{20}$$

where X is a $N \times N$ matrix given by

$$X_{ij} = \int_B dx \, \chi_i^*(x) \chi_j(x). \tag{21}$$

We call X overlap matrix. It is easy to show that the eigenvalues λ_i of the overlap matrix are in the range $0 \le \lambda_i \le 1$.

From the formula (20), we can find that the entropy has the upper bound[e] as

$$S_B(\psi) \le N \log 2. \tag{22}$$

The maximum entropy $N \log 2$ is proportional to the number of particles N, and thus follows an extensive property like thermal entropy. However,

---

[d]In this case, the projection Π_k in Eq. (3) is the projection to the subsector $\mathcal{H}_k^{(N)}$ in Eq. (13).
[e]We can also confirm that the classical part S_{cl} is bounded as $S_{cl}(\rho; A) \lesssim \mathcal{O}(\log N)$.[1]

this upper bound is too generic. We expect that EE for ground states is not extensive but sub-extensive in local models. In fact, we will see in the next section that the entropy of a ground state of N free fermions behaves as $S \sim \mathcal{O}(\log N)$ in the large N limit.

5. Entanglement for free fermions in a circle

We now apply the formula (20) to N free fermions in a circle with length L, i.e., the target space M is a circle. The Hamiltonian is given by $H = \sum_{i=1}^{N} \frac{p_i^2}{2m}$, and we consider its ground state. The one-body eigenfunctions are given by $\chi_n(x) = \frac{1}{\sqrt{L}} e^{\frac{2\pi i}{L} n x}$ where n are integers. Supposing that the total number of particles N is odd ($N = 2K+1$), the N-body wave function for the ground state is given by the Slater determinant as

$$\psi(x_1, \cdots, x_N) = \frac{1}{\sqrt{N!}} \sum_{\sigma \in S_N} \text{sgn}(\sigma) \chi_{-K}(x_{\sigma(1)}) \cdots \chi_K(x_{\sigma(N)}). \tag{23}$$

Thus, the target space entanglement for a subregion B can be obtained by the formula (20) with the $N \times N$ overlap matrix

$$X_{nn'} = \int_B dx\, \chi_n^*(x) \chi_{n'}(x), \tag{24}$$

where n, n' runs in $-K, \ldots, K$.

5.1. *Single interval*

In this subsection, we consider the case where the subregion B is a single interval I_1 in the circle. We parameterize the length of the interval as rL ($0 \leq r \leq 1$), i.e., r is the ratio of the interval to the circle.

In the large N limit, the asymptotic behavior of the entropy can be obtained as

$$S_{I_1} \sim \frac{1}{3} \log[2N \sin(\pi r)] + \Upsilon_1 \tag{25}$$

with

$$\Upsilon_1 = i \int_{-\infty}^{\infty} dw \frac{\pi w}{\cosh^2(\pi w)} \log \frac{\Gamma\left(\frac{1}{2} + iw\right)}{\Gamma\left(\frac{1}{2} - iw\right)} \sim 0.495018. \tag{26}$$

We show the plot of the entropy with the large N result (25) in Fig. 1. It shows that the entropy is sub-extensive (not proportional to N). Furthermore, the large N behavior Eq. (25) agrees with the EE for the single interval in $c = 1$ CFTs on the circle[17] if we regard N as a (dimensionless) cutoff.

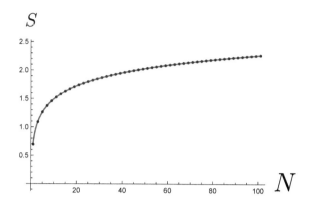

Fig. 1. EE for the half region. The red dots are the EE S for $N = 1, 3, \cdots, 101$. The blue curve represents the large N result (25) with $r = 1/2$.

5.2. *Entanglement entropy and mutual information for two intervals*

In this subsection, we consider two disjoint intervals I_1 and I_2 in the circle. Suppose that the coordinates of the circle is x moving in $-\frac{L}{2} \le x \le \frac{L}{2}$ with the periodic condition $x \sim x + L$. We take the two intervals as $I_1 = \left(-\frac{d+r}{2}L, -\frac{d-r}{2}L\right)$ and $I_2 = \left(\frac{d-r}{2}L, \frac{d+r}{2}L\right)$.

The EE for the subregion $I_1 \cup I_2$ can be analytically computed in the large N limit[f] as

$$S_{I_1 \cup I_2} \sim \frac{1}{3}\left[2\log[2N\sin(\pi r)] + \log\frac{\sin[\pi(d+r)]\sin[\pi(d-r)]}{\sin^2(\pi d)}\right] + 2\Upsilon_1.$$
(27)

We can also evaluated the target space mutual information;

$$I(I_1; I_2) := S(I_1) + S(I_2) - S(I_1 \cup I_2).$$
(28)

The large N behavior is

$$I(I_1; I_2) \sim \frac{1}{3}\log\frac{\sin^2(\pi d)}{\sin[\pi(d+r)]\sin[\pi(d-r)]}.$$
(29)

The mutual information is finite even in the large N limit. In addition, Eq. (29) agrees with the result in a $c = 1$ CFT (free compact boson at the self-dual radius[18]), although the reason is not understood well.

The plot of the target space mutual information (28) is Fig. 2.

---

[f]See Ref. 1.

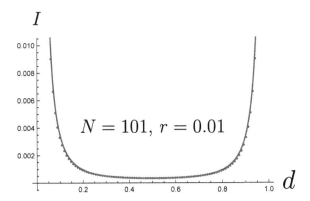

Fig. 2. Mutual information for two intervals. We take $N = 101$ and set the parameter r as $r = 0.01$ (length of each interval is rL). The red dots represent the mutual information for some values of d. The blue curve represents the large N result (29).

6. Brief conclusion

The algebraic approach is a powerful method of characterizing entanglement. This approach might be useful beyond the target space entanglement. A similar idea to define entropy based on observables is also investigated as the observational entropy (see, *e.g.*, Ref. 19).

We have used the algebraic approach to define the target space entanglement of particles. In particular, we consider non-interacting fermions, which can be regarded as the singlet sectors of one-matrix models. It is more interesting to consider entanglement in multi-matrix models, and its relation to holography.

Acknowledgments

SS thanks the organizers of East Asia Joint Symposium on Fields and Strings 2021 for the opportunity to present the talk at Osaka City University. SS acknowledges support from JSPS KAKENHI Grant Number JP 21K13927.

References

1. S. Sugishita, Target space entanglement in quantum mechanics of fermions and matrices, *JHEP* **08**, p. 046 (2021).
2. S. Ryu and T. Takayanagi, Holographic derivation of entanglement entropy from AdS/CFT, *Phys. Rev. Lett.* **96**, p. 181602 (2006).

3. E. A. Mazenc and D. Ranard, Target Space Entanglement Entropy (10 2019).
4. S. R. Das, A. Kaushal, G. Mandal and S. P. Trivedi, Bulk Entanglement Entropy and Matrices, *J. Phys. A* **53**, p. 444002 (2020).
5. S. R. Das, A. Kaushal, S. Liu, G. Mandal and S. P. Trivedi, Gauge Invariant Target Space Entanglement in D-Brane Holography (11 2020).
6. H. R. Hampapura, J. Harper and A. Lawrence, Target space entanglement in Matrix Models (12 2020).
7. A. Frenkel and S. A. Hartnoll, Entanglement in the Quantum Hall Matrix Model (11 2021).
8. A. Tsuchiya and K. Yamashiro, Target space entanglement in a matrix model for the bubbling geometry (1 2022).
9. S. R. Das, S. Hampton and S. Liu, Entanglement Entropy and Phase Space Density: Lowest Landau Levels and 1/2 BPS states (1 2022).
10. S. R. Das, Geometric entropy of nonrelativistic fermions and two-dimensional strings, *Phys. Rev. D* **51**, 6901 (1995).
11. S. A. Hartnoll and E. Mazenc, Entanglement entropy in two dimensional string theory, *Phys. Rev. Lett.* **115**, p. 121602 (2015).
12. M. Ohya and D. Petz, *Quantum entropy and its use* (Springer Science & Business Media, 2004).
13. H. Casini, M. Huerta and J. A. Rosabal, Remarks on entanglement entropy for gauge fields, *Phys. Rev. D* **89**, p. 085012 (2014).
14. D. Harlow, The Ryu–Takayanagi Formula from Quantum Error Correction, *Commun. Math. Phys.* **354**, 865 (2017).
15. M. Goldstein and E. Sela, Symmetry-resolved entanglement in many-body systems, *Phys. Rev. Lett.* **120**, p. 200602 (2018).
16. R. Bonsignori, P. Ruggiero and P. Calabrese, Symmetry resolved entanglement in free fermionic systems, *J. Phys. A* **52**, p. 475302 (2019).
17. P. Calabrese and J. L. Cardy, Entanglement entropy and quantum field theory, *J. Stat. Mech.* **0406**, p. P06002 (2004).
18. P. Calabrese, J. Cardy and E. Tonni, Entanglement entropy of two disjoint intervals in conformal field theory, *J. Stat. Mech.* **0911**, p. P11001 (2009).
19. D. Šafránek, A. Aguirre, J. Schindler and J. M. Deutsch, A Brief Introduction to Observational Entropy, *Found. Phys.* **51**, p. 101 (2021).

Jacobi Forms on Even Positive Definite Unimodular Lattices

K. Sun

Korea Institute for Advanced Study,
Seoul, 02455, South Korea
* E-mail: ksun@kias.re.kr

We study Jacobi forms associated to even positive definite unimodular lattices, in particular E_8 lattice and the Leech lattice. We require the Jacobi forms to be invariant under the orthogonal group of the lattice, in particular E_8 Weyl group and Conway group Co_0. These objects are of interest in both number theory and string theory, for example elliptic genera of E-strings, Monster CFT and so on. We establish the explicit structure theorem and determine the generators for the E_8 case with index $t \leq 13$ and the Leech case with index $t \leq 3$. This is a short summary of two joint papers [1,2] with Haowu Wang.

Keywords: Jacobi forms; Even positive definite unimodular lattice; E_8 lattice; Leech lattice; Conway group.

1. Introduction

Jacobi forms are some fundamental objects in number theory and also basic tools in string theory whenever a torus T^2 and a global symmetry is involved. Originally, Jacobi forms are introduced by Eichler and Zagier in their monograph [3]. These forms are holomorphic functions in two variables $(\tau, z) \in \mathbb{H} \times \mathbb{C}$ which are modular in τ with respective to $\mathrm{SL}_2(\mathbb{Z})$ and quasi-periodic in z. Later, Gritsenko [4] defined Jacobi forms of lattice index by replacing z with many variables associated with an integral positive definite lattice. The Jacobi form creates an elegant bridge between different types of modular forms. For example, Jacobi forms can be identified as vector valued modular forms through the theta decomposition. Jacobi forms also have many applications in mathematical physics, such as the elliptic genera of $K3$ surface, elliptic genera of 6d $(1,0)$ SCFTs and the topological string partition functions on various Calabi–Yau threefolds. It is a natural question to determine the structure of the space of Jacobi forms. This question was solved by Wirthmüller [5] for Jacobi forms associated with root systems not of E_8 type, where there is polynomial ring structure. For the E_8 root system, the structure becomes much more complicated due to the even unimodular property. In particular, the space is *not* a polynomial ring. This was studied before in[6-8], but the structure is still not explicit. Besides, little is known about spaces of Jacobi forms associated with other irreducible lattices of large rank. We aim to tackle these problems, in particular the E_8 lattice and the Leech lattice.

2. Jacobi Forms of Lattice Index

We review the basic notions of Jacobi forms of lattice index. Let L denote an even positive definite unimodular lattice equipped with bilinear form $(-,-)$. Here the unimodular means the lattice has determinant 1 and the even means all vectors have even norms. It is well-known such L can only exist in dimension $8k, k \in \mathbb{Z}$.

Definition 2.1. Let $k \in \mathbb{Z}$ be an integer and $t \in \mathbb{N}$ be a non-negative integer. If a holomorphic function $\varphi : \mathbb{H} \times (L \otimes \mathbb{C}) \to \mathbb{C}$ satisfies the conditions

(i) Quasi-periodicity:

$$\varphi(\tau, \mathfrak{z} + x\tau + y) = \exp\left(-t\pi i(x,x)\tau - 2t\pi i(x,\mathfrak{z})\right) \varphi(\tau, \mathfrak{z}), \quad x, y \in L,$$

(ii) Modularity: for $a, b, c, d \in \mathbb{Z}$ and $ad - bc = 1$,

$$\varphi\left(\frac{a\tau + b}{c\tau + d}, \frac{\mathfrak{z}}{c\tau + d}\right) = (c\tau + d)^k \exp\left(t\pi i \frac{c(\mathfrak{z}, \mathfrak{z})}{c\tau + d}\right) \varphi(\tau, \mathfrak{z}),$$

and the Fourier expansion of φ takes the form

$$\varphi(\tau, \mathfrak{z}) = \sum_{n=0}^{\infty} \sum_{\ell \in L} f(n, \ell) q^n \zeta^\ell, \quad q = e^{2\pi i \tau}, \quad \zeta^\ell = e^{2\pi i(\ell, \mathfrak{z})},$$

then it is called a *weak Jacobi form* of weight k and index t. If φ further satisfies that $f(n, \ell) = 0$ whenever $2nt - (\ell, \ell) < 0$, then it is called a *holomorphic Jacobi form*. If φ is invariant under the orthogonal group of lattice $\mathrm{O}(L)$, then it is called a $\mathrm{O}(L)$ *invariant Jacobi form*. Here $\mathrm{O}(L)$ contains all automorphism of L that keeps the bilinear form.

As L is unimodular, the theta decomposition (see[9] Corollary 2.6)) yields that every $\mathrm{O}(L)$ invariant weak Jacobi form of weight k and index 1 is a holomorphic Jacobi form and can be expressed as $g(\tau)\Theta_L(\tau, \mathfrak{z})$, where $g(\tau)$ is a modular form of weight $k - \frac{1}{2}\mathrm{rk}(L)$ on $\mathrm{SL}_2(\mathbb{Z})$, and $\Theta_L(\tau, \mathfrak{z})$ is the Jacobi theta function of L defined by

$$\Theta_L(\tau, \mathfrak{z}) = \sum_{\ell \in L} e^{\pi i(\ell, \ell)\tau + 2\pi i(\ell, \mathfrak{z})}.$$

For index $t > 1$, the structure of $O(L)$ invariant Jacobi forms becomes nontrivial. The cases of interest are those with large $O(L)$, i.e. L is highly symmetrical, then the space of $O(L)$ invariant Jacobi forms will be relatively simple.

We focus on two most interesting cases. The first is the E_8 root lattice, where we require the Jacobi forms to be E_8 Weyl invariant:

$$\varphi(\tau, \sigma(\mathfrak{z})) = \varphi(\tau, \mathfrak{z}), \quad \sigma \in W(E_8).$$

E_8 Weyl invariant Jacobi forms are frequently used in string theory, for example in E-string theory, which is the simplest six dimensional $(1,0)$ superconformal field theory with manifest E_8 symmetry[6,10,11]. Due to this importance, it is highly interesting to determine the precise structure of space of E_8 Weyl invariant Jacobi forms to high index.

The second case is the Leech lattice, which is the unique even positive definite unimodular lattice in dimension 24 that has no roots. It was discovered by Leech[12], and its uniqueness was proved by Conway[13]. This lattice has many remarkable properties. For example, it plays a role in constructing the fake monster Lie algebra[14] and proving the monstrous moonshine conjecture[15], and it achieves the densest sphere packing in 24 dimension[16]. The group Co_0 is the automorphism group of the Leech lattice, whose structure was first described by Conway[17]. It is natural the require the Jacobi forms on the Leech lattice to be *Conway invariant*:

$$\varphi(\tau, \sigma(\mathfrak{z})) = \varphi(\tau, \mathfrak{z}), \quad \sigma \in Co_0.$$

The quotient of Co_0 by its center gives a sporadic simple group of order $4,157,776,806,543,360,000$. Therefore, the Leech lattice is highly symmetrical, and we expect that the space of Conway invariant Jacobi forms will not be too large. Due to the importance of the Leech lattice and the Conway group, we also expect that Conway invariant Jacobi forms will have some applications in mathematics and physics. These motivate us to study such Jacobi forms.

3. E_8 Weyl Invariant Jacobi Forms

Denote the vector spaces of $W(E_8)$-invariant weak and holomorphic Jacobi forms of weight k and index t respectively by $J_{k,E_8,t}^{w,W(E_8)} \supsetneq J_{k,E_8,t}^{W(E_8)}$. Let $M_*(\mathrm{SL}_2(\mathbb{Z})) = \mathbb{C}[E_4, E_6]$ be the ring of modular forms on $\mathrm{SL}_2(\mathbb{Z})$. For fixed index t, these are free $M_*(\mathrm{SL}_2(\mathbb{Z}))$-modules

$$J_{*,E_8,t}^{w,W(E_8)} := \bigoplus_{k\in\mathbb{Z}} J_{k,E_8,t}^{w,W(E_8)}, \quad J_{*,E_8,t}^{W(E_8)} := \bigoplus_{k\in\mathbb{Z}} J_{k,E_8,t}^{W(E_8)}.$$

We aim to describe the explicit structure of the above modules using Sakai's forms A_1, A_2, B_2, A_3, B_3, A_4, B_4, A_5, B_6. These were some algebraically independent $W(E_8)$-invariant holomorphic Jacobi forms first constructed by Sakai in the study on E-strings[11]. Due to its importance, we briefly explain how Sakai constructed these forms. One starts with the E_8 Jacobi theta function $\vartheta_{E_8}(\tau, \mathfrak{z})$ which is the unique $W(E_8)$-invariant holomorphic Jacobi form of weight 4 and index 1. Acting the index raising Hecke operators $T_-(t)$ on ϑ_{E_8}, one obtains $W(E_8)$-invariant holomorphic Jacobi forms of weight 4 and arbitrary index t: $X_t(\tau, \mathfrak{z}) = 1 + O(q)$. Sakai's forms A_j are constructed as

$$A_j(\tau, \mathfrak{z}) = X_j(\tau, \mathfrak{z}), \ j = 1, 2, 3, 5, \quad A_4(\tau, \mathfrak{z}) = \vartheta_{E_8}(\tau, 2\mathfrak{z}).$$

To construct B_t, one first takes an appropriate modular form g_t of weight 2 on the congruence subgroup $\Gamma_0(t)$ of $SL_2(\mathbb{Z})$. Then the trace sum of $g_t(\tau)\vartheta_{E_8}(t\tau, t\mathfrak{z})$ with respect to the cosets of $\Gamma_0(t) \backslash SL_2(\mathbb{Z})$ defines a $W(E_8)$-invariant holomorphic Jacobi form of weight 6 and index t. That is the desired B_t.

Rather surprisingly, Del Zotto, Gu, Huang, Kashani-Poor, Klemm and Lockhart[18] discovered an exceptional $W(E_8)$-invariant holomorphic Jacobi form of weight 16 and index 5 defined by the polynomial $P_{16,5}$:

$$864A_1^3A_2 + 3825A_1B_2^2 - 770A_3B_2E_6 - 840A_2B_3E_6 + 60A_1B_4E_6 + 21A_5E_6^2.$$

They checked numerically that $P_{16,5}$ vanishes at the zero points of E_4 for general lattice variable $\mathfrak{z}$ and then conjectured that the quotient $P_{16,5}/E_4$ is holomorphic. They did not find other similar polynomials, so they further conjectured that any Jacobi form expressed as a polynomial in A_i, B_j and E_6 which vanishes at the zero points of E_4 must be divisible by the above polynomial. In[1] we proved their conjectures. Utilizing this distinguished Jacobi form $P_{16,5}/E_4$, we are able to give a full description of $W(E_8)$-invariant Jacobi forms of arbitrary index in terms of Sakai's forms.

Theorem 3.1.

(1) The quotient $P_{16,5}/E_4$ is a $W(E_8)$-invariant holomorphic Jacobi form of weight 12 and index 5.

(2) For any $W(E_8)$-invariant Jacobi form $P \in \mathbb{C}[E_6, A_i, B_j]$, if P/E_4 is holomorphic on $\mathbb{H} \times (E_8 \otimes \mathbb{C})$, then

$$\frac{P}{P_{16,5}} \in \mathbb{C}[E_6, A_i, B_j].$$

(3) Every $W(E_8)$-invariant weak Jacobi form of index t can be expressed uniquely as

$$\frac{\sum_{j=0}^{t_1} P_j E_4^j P_{16,5}^{t_1-j}}{\Delta^{N_t} E_4^{t_1}},$$

where

(i) t_1 is the integer part of $t/5$;
(ii) $P_{t_1} \in \mathbb{C}[E_4, E_6, A_i, B_j]$;
(iii) $P_j \in \mathbb{C}[E_6, A_i, B_j]$ for $0 \le j < t_1$;
(iv) N_t is defined as follows

$$N_t = \begin{cases} 5t_0, & \text{if } t = 6t_0 \text{ or } 6t_0 + 1, \\ 5t_0 + 1, & \text{if } t = 6t_0 + 2, \\ 5t_0 + 2, & \text{if } t = 6t_0 + 3, \\ 5t_0 + 3, & \text{if } t = 6t_0 + 4 \text{ or } 6t_0 + 5. \end{cases}$$

To determine the explicit generators for each index, we first calculate the Fourier expansions of Sakai's forms A_i and B_j up to q^9-terms. The Fourier expansions involve 268 Weyl orbits of vectors of norm $\frac{1}{2}(v, v) \le 54$. We then express these Weyl orbits as polynomials in the eight fundamental Weyl orbits. Using these data and the above theorem, we successfully determine all generators of $J_{*, E_8, t}^{\mathrm{w}, W(E_8)}$ for $1 \le t \le 13$ and following generating series for the space dimensions.

Theorem 3.2. *Let $d_{k,t}$ denote the number of generators of weight k of $J_{*, E_8, t}^{\mathrm{w}, W(E_8)}$. For $1 \le t \le 13$ the Laurent polynomials*

$$P_t^{\mathrm{w}} := \sum_{k \in \mathbb{Z}} d_{k,t} x^k$$

describing the weights of generators are determined as follows

$$P_1^{\mathrm{w}} = x^4, \qquad P_2^{\mathrm{w}} = x^{-4} + x^{-2} + 1, \qquad P_3^{\mathrm{w}} = x^{-8} + x^{-6} + x^{-4} + x^{-2} + 1,$$

$$P_4^{\mathrm{w}} = x^{-16} + x^{-14} + x^{-12} + x^{-10} + 2x^{-8} + x^{-6} + x^{-4} + x^{-2} + 1,$$

$$P_5^{\mathrm{w}} = 2x^{-16} + 2x^{-14} + 3x^{-12} + 2x^{-10} + 2x^{-8} + x^{-6} + x^{-4} + x^{-2} + 1,$$

$$\cdots$$

$$P_{13}^{\mathrm{w}} = 2x^{-52} + 10x^{-50} + 24x^{-48} + 32x^{-46} + 37x^{-44} + 28x^{-42} + 29x^{-40}$$
$$+ 28x^{-38} + 26x^{-36} + 23x^{-34} + 22x^{-32} + 18x^{-30} + 16x^{-28} + 14x^{-26}$$
$$+ 12x^{-24} + 9x^{-22} + 8x^{-20} + 6x^{-18} + 5x^{-16} + 4x^{-14} + 3x^{-12}$$
$$+ 2x^{-10} + 2x^{-8} + x^{-6} + x^{-4} + x^{-2} + 1.$$

The Laurent expansion of the following rational function at $x = 0$ gives the dimension of the space of weak Jacobi forms of arbitrary weight and given index t

$$\frac{P_t^{\mathrm{w}}}{(1 - x^4)(1 - x^6)} = \frac{\sum_{k \in \mathbb{Z}} d_{k,t} x^k}{(1 - x^4)(1 - x^6)} = \sum_{k \in \mathbb{Z}} \dim J_{k,E_8,t}^{\mathrm{w},W(E_8)} x^k.$$

We further compute the dimension of the space of $W(E_8)$-invariant holomorphic Jacobi forms of weight 4 and small index and construct explicit generators. These so-called holomorphic Jacobi forms of singular (i.e. possible minimal positive) weight are usually difficult to determine and construct in the theory of modular forms.

Proposition 3.1. *The dimension of the space $J_{4,E_8,t}^{W(E_8)}$ for $t \leq 11$ is formulated in Table 3.1.*

Table 1. The dimension of $J_{4,E_8,t}^{W(E_8)}$

t	1	2	3	4	5	6	7	8	9	10	11
dim	1	1	1	2	1	1	2	2	2	2	2

4. Conway Invariant Jacobi Forms on the Leech Lattice

Let $J_{*,\Lambda,t}^{\mathrm{w},\mathrm{Co_0}}$ and $J_{*,\Lambda,t}^{\mathrm{Co_0}}$ denote the spaces of Conway invariant weak and holomorphic Jacobi forms of integral weight and given index t respectively. The following is our main theorem[2].

Theorem 4.1. *As free modules over $M_*(\mathrm{SL}_2(\mathbb{Z}))$,*

(1) $J_{*,\Lambda,2}^{\mathrm{w},\mathrm{Co_0}}$ is generated by four forms of weights -4, -2, 0, 0.

(2) $J_{*,\Lambda,2}^{\mathrm{Co_0}}$ is generated by four forms of weights 12, 12, 14, 16.

(3) $J_{*,\Lambda,3}^{\mathrm{w},\mathrm{Co_0}}$ is generated by ten forms of weights -14, -12, -12, -12, -10, -8, -6, -4, -2, 0.

(4) $J_{*,\Lambda,3}^{\mathrm{Co_0}}$ is generated by ten forms of weights 12, 12, 12, 14, 14, 16, 16, 16, 18, 18.

To prove the above theorem, we first use the differential operators approach in[7] to estimate the minimal weight of weak Jacobi forms of a given index. Then we combine the arguments in[1,7,11] to construct generators such as Hecke operator. We also construct one of the singular-weight generators of $J^{\mathrm{Co}_0}_{*,\Lambda,t}$ as the t-th Fourier–Jacobi coefficient of Borcherds' automorphic form Φ_{12} for the unimodular lattice $\mathrm{II}_{26,2}$ (see[19]). The main difficulty of the proof is to calculate the Fourier expansions of generators, because Conway invariant Jacobi forms have unwieldy Fourier expansions in 25 variables. To overcome this difficulty, we write the Fourier expansion of a Jacobi form in terms of *Conway orbits* defined as the Co_0-invariant exponential polynomials

$$\mathrm{orb}(v) = \sum_{\sigma \in \mathrm{Co}_0/(\mathrm{Co}_0)_v} e^{2\pi i(\sigma(v),\mathfrak{z})},$$

where $v \in \Lambda$ and $(\mathrm{Co}_0)_v$ is the stabilizer of Co_0 with respect to v. The Conway orbits $\mathrm{orb}(v)$ of type $\frac{1}{2}(v,v) \le 16$ are available in[20]. In order to calculate the Fourier expansions of products of Jacobi forms, we have to know the decomposition of some products $\mathrm{orb}(v)\,\mathrm{orb}(u)$ into linear combinations of Conway orbits. We determine such non-trivial decompositions by comparing the Fourier–Jacobi expansion of Φ_{12} and the Borcherds denominator formula for the fake monster Lie algebra (see[14,19]). Combining these arguments together, we prove the theorem.

The differential operator H also called *heat operator* used in the proof and the construction of generators is defined by the following lemma.

Lemma 4.1. *Given a Conway invariant weak Jacobi form of weight k and index $t \ge 1$*

$$\varphi(\tau,\mathfrak{z}) = \sum_{n=0}^{\infty} \sum_{r \in \Lambda/\mathrm{Co}_0} f(n,r)q^n \cdot \mathrm{orb}(r).$$

Then $H_k(\varphi)$ is a Conway invariant weak Jacobi form of weight $k+2$ and index t, where

$$H_k(\varphi)(\tau,\mathfrak{z}) = \mathcal{H}(\varphi)(\tau,\mathfrak{z}) + \frac{12-k}{12}E_2(\tau)\varphi(\tau,\mathfrak{z}),$$

$$\mathcal{H}(\varphi)(\tau,\mathfrak{z}) = \sum_{n\in\mathbb{N}} \sum_{r\in\Lambda/\mathrm{Co}_0} \left(n - \frac{(r,r)}{2t}\right) f(n,r)q^n \cdot \mathrm{orb}(r),$$

and $E_2(\tau) = 1 - 24\sum_{n\ge 1}\sigma(n)q^n$ is the Eisenstein series of weight 2.

Similar to Sakai's A_i and B_j forms for the E_8 case, we can also define analogously Conway invariant holomorphic Jacobi forms A_i and B_j for the Leech lattice which have weight 12 and 14 respectively. The other type of Conway invariant holomorphic Jacobi forms of singular weight 12 come from the famous Borcherds automorphic form Φ_{12} given by the *Weyl–Kac–Borcherds denominator identity* of the fake monster Lie algebra. In terms of Conway invariant Jacobi forms, one can express Φ_{12} as follows (see [19,21])

$$\Phi_{12}(Z) = \Delta(\tau) \cdot \exp\left(-\sum_{m=1}^{\infty} ((\Delta^{-1}A_1)|T_-(m))(\tau,\mathfrak{z})e^{2\pi i m\omega} \right)$$

$$= \sum_{m=0}^{\infty} \Phi_{12,m}(\tau,\mathfrak{z})e^{2\pi i m\omega},$$

where $T_-(m)$ are Hecke operators. Then $\Phi_{12,m}(\tau,\mathfrak{z})$ is a Conway invariant holomorphic Jacobi forms of singular weight 12 and index m. The above operator and basic forms allow us to construct the explicit generators for the space of holomorphic forms at index 2 and 3.

Theorem 4.2. *The free* $\mathbb{C}[E_4, E_6]$*-module* $J_{*,\Lambda,2}^{\mathrm{Co}_0}$ *is generated by* A_2, $\Phi_{12,2}$, B_2 *and* HB_2, *which have weights* 12, 12, 14, 16 *respectively.*

Theorem 4.3. *The free* $\mathbb{C}[E_4, E_6]$*-module* $J_{*,\Lambda,3}^{\mathrm{Co}_0}$ *is generated by ten forms*

$$A_3, \ \Phi_{12,3}, \ \Psi_{12,3}, \ B_3, \ \Psi_{14,3}, \ HB_3, \ H\Psi_{14,3}, \ \Psi_{16,3}, \ H^2B_3, \ H^2\Psi_{14,3}$$

which have weights 12, 12, 12, 14, 14, 16, 16, 16, 18, 18 *respectively.*

For the explicit construction of holomorphic forms $\Psi_{12,3}$, $\Psi_{14,3}$ and $\Psi_{16,3}$, we refer to [2]. We also use these holomorphic generators to construct the weak generators [2].

As applications, we determine many product decompositions of Conway orbits by means of modular linear relations among Conway invariant holomorphic Jacobi forms. These results are formulated in Appendix A of [2]. For example,

$$O_2 \otimes O_2 = 1965600 O_0 \oplus 4600 O_2 \oplus 552 O_3 \oplus 46 O_4 \oplus 2 O_5 \oplus 2 O_{6b} \oplus O_{8a},$$

$$O_2 \otimes O_3 = 47104 O_2 \oplus 11178 O_3 \oplus 2048 O_4 \oplus 275 O_5 \oplus 24 O_{6a} \oplus O_7 \oplus O_{8c}.$$

Here the Conway orbits follow the notion of the ALTAS [20]. It would be very difficult to compute these product decompositions in a brutal way due to the huge size of the Conway orbits. We also classify Conway invariant holomorphic Jacobi forms of singular weight 12 and index $t \leq 3$ with non-trivial character and use the Fourier expansions of our Jacobi forms to determine all conjugate relations among Conway orbits of type $\frac{1}{2}(v,v) \leq 16$ modulo 2Λ and 3Λ. Besides, we calculate the pullbacks of Conway invariant Jacobi forms and Conway orbits along Leech vectors of types 2, 3 and 4, which characterize the intersection behaviors of Leech vectors.

We know from Borcherds' thesis [22] that $\Lambda/4\Lambda$ have 31 orbits with respect to Co_0. We give an explicit description of the representative system of minimal length of $\Lambda/4\Lambda$ in [2]. Borcherds' result yields that the rank of $J_{*,\Lambda,4}^{w,Co_0}$ is 31. It seems very difficult to determine and construct the associated 31 generators of index 4.

The famous 2d Monster CFT has one single character $\chi = \Theta_\Lambda(\tau)/\eta^{24} - 24$ where $\Theta_L(\tau) = \Theta_L(\tau, \mathfrak{z} = 0)$. This can be seen as the index one case of the discussion in this section. It is interesting to consider whether there exist certain level two generalization of 2d Monster CFT which has 4 characters, and level three generalization which has 10 characters, and so on.

References

1. K. Sun and H. Wang, Weyl invariant E_8 Jacobi forms and E-strings, Preprint, (2021).

2. K. Sun and H. Wang, Conway invariant Jacobi forms on the Leech lattice, Preprint, (2021).

3. M. Eichler and D. Zagier, *The theory of Jacobi forms*, Progress in Mathematics, Vol. 55 (Birkhäuser Boston, Inc., Boston, MA, 1985).

4. V. A. Gritsenko, Fourier-Jacobi functions in n variables, *Zap. Nauchn. Sem. Leningrad. Otdel. Mat. Inst. Steklov. (LOMI)* **168**, 32 (1988).

5. K. Wirthmüller, Root systems and Jacobi forms, *Compositio Math.* **82**, 293 (1992).

6. K. Sakai, E_n Jacobi forms and Seiberg-Witten curves, *Commun. Number Theory Phys.* **13**, 53 (2019).

7. H. Wang, Weyl invariant E_8 Jacobi forms, *Commun. Number Theory Phys.* **15**, 517 (2021).

8. H. Wang, Weyl invariant Jacobi forms: a new approach, *Adv. Math.* **384**, Paper No. 107752, 13 (2021).

9. V. Gritsenko, Modular forms and moduli spaces of abelian and $K3$ surfaces, *Algebra i Analiz* **6**, 65 (1994).

10. J. A. Minahan, D. Nemeschansky, C. Vafa and N. P. Warner, E strings and N=4 topological Yang-Mills theories, *Nucl. Phys. B* **527**, 581 (1998).

11. K. Sakai, Topological string amplitudes for the local $\frac{1}{2}$K3 surface, *PTEP. Prog. Theor. Exp. Phys.* , 033B09, 29 (2017).

12. J. Leech, Notes on sphere packings, *Canadian J. Math.* **19**, 251 (1967).

13. J. H. Conway, A characterisation of Leech's lattice, *Invent. Math.* **7**, 137 (1969).

14. R. E. Borcherds, The monster Lie algebra, *Adv. Math.* **83**, 30 (1990).
15. R. E. Borcherds, Monstrous moonshine and monstrous Lie superalgebras, *Invent. Math.* **109**, 405 (1992).
16. H. Cohn, A. Kumar, S. D. Miller, D. Radchenko and M. Viazovska, The sphere packing problem in dimension 24, *Ann. of Math. (2)* **185**, 1017 (2017).
17. J. H. Conway, A group of order $8,315,553,613,086,720,000$, *Bull. London Math. Soc.* **1**, 79 (1969).
18. M. Del Zotto, J. Gu, M.-X. Huang, A.-K. Kashani-Poor, A. Klemm and G. Lockhart, Topological Strings on Singular Elliptic Calabi-Yau 3-folds and Minimal 6d SCFTs, *JHEP* **03**, p. 156 (2018).
19. R. E. Borcherds, Automorphic forms on $O_{s+2,2}(\mathbf{R})$ and infinite products, *Invent. Math.* **120**, 161 (1995).
20. J. H. Conway, R. T. Curtis, S. P. Norton, R. A. Parker and R. A. Wilson, ATLAS *of finite groups* (Oxford University Press, Eynsham, 1985), Maximal subgroups and ordinary characters for simple groups, With computational assistance from J. G. Thackray.
21. V. A. Gritsenko, Reflective modular forms and their applications, *Uspekhi Mat. Nauk* **73**, 53 (2018).
22. R. E. Borcherds, The leech lattice and other lattices, Ph.D. dissertation, Cambridge Univ., Cambridge, (1985).

Printed in the United States
by Baker & Taylor Publisher Services